Universitext

Editors

J. Ewing
F.W. Gehring
P.R. Halmos

Universitext

Editors: J. Ewing, F.W. Gehring, and P.R. Halmos

Booss/Bleecker: Topology and Analysis
Charlap: Bieberbach Groups and Flat Manifolds
Chern: Complex Manifolds Without Potential Theory
Chorin/Marsden: A Mathematical Introduction to Fluid Mechanics
Cohn: A Classical Invitation to Algebraic Numbers and Class Fields
Curtis: Matrix Groups, 2nd ed.
van Dalen: Logic and Structure
Devlin: Fundamentals of Contemporary Set Theory
Edwards: A Formal Background to Mathematics I a/b
Edwards: A Formal Background to Mathematics II a/b
Endler: Valuation Theory
Frauenthal: Mathematical Modeling in Epidemiology
Gardiner: A First Course in Group Theory
Gårding/Tambour: Algebra for Computer Science
Godbillon: Dynamical Systems on Surfaces
Greub: Multilinear Algebra
Hermes: Introduction to Mathematical Logic
Humi/Miller: Second Order Ordinary Differential Equations
Hurwitz/Kritikos: Lectures on Number Theory
Kelly/Matthews: The Non-Euclidean, The Hyperbolic Plane
Kostrikin: Introduction to Algebra
Luecking/Rubel: Complex Analysis: A Functional Analysis Approach
Lu: Singularity Theory and an Introduction to Catastrophe Theory
Marcus: Number Fields
McCarthy: Introduction to Arithmetical Functions
Mines/Richman/Ruitenburg: A Course in Constructive Algebra
Meyer: Essential Mathematics for Applied Fields
Moise: Introductory Problem Course in Analysis and Topology
Øksendal: Stochastic Differential Equations
Porter/Woods: Extensions of Hausdorff Spaces
Rees: Notes on Geometry
Reisel: Elementary Theory of Metric Spaces
Rey: Introduction to Robust and Quasi-Robust Statistical Methods
Rickart: Natural Function Algebras
Smith: Power Series From a Computational Point of View
Smoryński: Self-Reference and Modal Logic
Stanisić: The Mathematical Theory of Turbulence
Stroock: An Introduction to the Theory of Large Deviations
Sunder: An Invitation to von Neumann Algebras
Tolle: Optimization Methods
Tondeur: Foliations on Riemannian Manifolds

Lars Gårding   Torbjörn Tambour

Algebra for Computer Science

Springer-Verlag
New York Berlin Heidelberg London Paris Tokyo

Lars Gårding
Torbjörn Tambour
Department of Mathematics
Lunds Universitets Matematiska Institution
S-22100 Lund
Sweden

With 6 Illustrations

Mathematics Subject Classification (1980): 02/20, 10/30

Library of Congress Cataloging-in-Publication Data
Gårding, Lars,
  Algebra for computer science.
  (Universitext)
  Bibliography: p.
  Includes index.
  1. Electronic data processing—Mathematics
I. Tambour, Torbjörn. II. Title.
QA76.9.M35G37 1988   004'.01'512   88-15997

© 1988 by Springer-Verlag New York Inc.
All rights reserved. This work may not be translated or copied in whole or in part without the written permission of the publisher (Springer-Verlag, 175 Fifth Avenue, New York, New York 10010, USA), except for brief excerpts in connection with reviews or scholarly analysis. Use in connection with any form of information storage and retrieval, electronic adaptation, computer software, or by similar or dissimilar methodology now known or hereafter developed is forbidden.
The use of general descriptive names, trade names, trademarks, etc. in this publication, even if the former are not especially identified, is not to be taken as a sign that such names, as understood by the Trade Marks and Merchandise Marks Act, may accordingly be used freely by anyone.

Camera-ready copy provided by the authors.
Printed and bound by R.R. Donnelley and Sons, Harrisonburg, Virginia.
Printed in the United States of America.

9 8 7 6 5 4 3 2 1

ISBN 0-387-96780-X Springer-Verlag New York Berlin Heidelberg
ISBN 3-540-96780-X Springer-Verlag Berlin Heidelberg New York

Preface

The aim of this book is to give the reader a general education in number theory, algebra and group theory and in the application of these parts of mathematics to computer science. The only prerequisite is knowledge of the elements of linear algebra and a certain mathematical maturity.

The wide range covered - number theory, abstract algebra, modules, rings and fields, groups, Boolean algebra and automata - would not have been feasible if some parts of basic theory had not been left to the reader in the form of straight-forward exercises labelled R for Reader and referred to as proven results. Some of them serve to shorten some of the more barren stretches of definitions and simple properties of algebra. In others the reader is supposed to reap the fruits of mathematical theory himself. The authors hope that the serious purpose of these exercises will motivate the reader to work them through. The book also contains many ordinary exercises.

The sections on computer science cover classical subjects like the cost of computing, pseudo-random numbers, the fast Fourier transform, algebraic complexity theory, shift registers and coding, counting of Boolean functions and the characterization of languages accepted by finite automata.

We are indebted to Anders Melin for trying out our first draft in an algebra course. Dan Laksov and Johan Håstad read later drafts and saved us from many blunders. We also thank Nils Dencker for putting together (*larsg.ty*), a typesetting program selected from TEX.

Lars Gårding *Torbjörn Tambour*

Contents

Chapter 1 **Number theory** 1

1.1 Divisibility 1
1.2 Congruences 5
1.3 The theorems of Fermat, Euler and Wilson 7
1.4 Squares and the quadratic reciprocity theorem 11
1.5 The Gaussian integers 14
1.6 Algebraic numbers 16
1.7 Appendix. Primitive elements and a theorem by Gauss 18
Literature 20

Chapter 2 **Number theory and computing** 21

2.1 The cost of arithmetic operations 21
2.2 Primes and factoring 25
2.3 Pseudo-random numbers 29
Literature 33

Chapter 3 **Abstract algebra and modules** 34

3.1 The four operations of arithmetic 34
3.2 Modules 38
3.3 Module morphisms. Kernels and images 43
3.4 The structure of finite modules 46
3.5 Appendix. Finitely generated modules 51
Literature 52

Chapter 4 **The finite Fourier transform** 53

4.1 Characters of modules 53
4.2 The finite Fourier transform 54
4.3 The finite Fourier transform and the quadratic reciprocity law ... 58
4.4 The fast Fourier transform 60
Literature 69

Chapter 5 **Rings and fields** 70

5.1 Definitions and simple examples 70
5.2 Modules over a ring. Ideals and morphisms 78
5.3 Abstract linear algebra 84

Literature 90

Chapter 6 Algebraic complexity theory 91

6.1 Polynomial rings in several variables 91
6.2 Complexity with respect to multiplication 95
6.3 Appendix. The fast Fourier transform is optimal 102
Literature 106

Chapter 7 Polynomial rings, algebraic fields, finite fields 107

7.1 Divisibility in a polynomial ring 107
7.2 Algebraic numbers and algebraic fields 115
7.3 Finite fields 123
Literature 125

Chapter 8 Shift registers and coding 126

8.1 The theory of shift registers 126
8.2 Generalities about coding 130
8.3 Cyclic codes 132
8.4 The BCH codes and the Reed-Solomon codes 135
8.5 Restrictions for error-correcting codes 137
Literature 140

Chapter 9 Groups 141

9.1 General theory 141

9.1.1 Groups and subgroups 141
9.1.2 Groups of bijections and normal subgroups 144
9.1.3 Groups acting on sets 145

9.2 Finite groups 150

9.2.1 Counting elements 150
9.2.2 Symmetry groups and the dihedral groups 152
9.2.3 The symmetric and alternating groups 154
9.2.4 Groups of low order 158
9.2.5 Applications of group theory to combinatorics 162
Literature 167

Chapter 10 **Boolean algebra** 168

10.1 Boolean algebras and rings 168
10.2 Finite Boolean algebras 172
10.3 Equivalence classes of switching functions 175
Literature 181

Chapter 11 **Monoids, automata, languages** 182

11.1 Matrices with elements in a non-commutative algebra 182
11.2 Monoids and languages 183
11.3 Automata and rational languages 186
11.4 Every rational language is accepted by a finite automaton 188
Literature 191

References 192
Index 194

CHAPTER 1

Number theory

Number theory is the oldest branch of mathematics. In fact, counting was a necessity of even very primitive life. Counting without regard for the nature of the objects counted gave us the natural numbers

$$\mathbf{N} = \{1, 2, \ldots\},$$

the first mathematical model. In this chapter we shall study the simplest properties of divisibility, the primes, the Gaussian integers, and the algebraic numbers.

1.1 Divisibility

When a, b, c are natural numbers and $a = bc$, we say that b and c divide a or are *divisors* of a and that a is a *multiple* of b and c. That b divides a is sometimes written as $b|a$.

Every natural number > 1 has at least two divisors, namely 1 and the number itself. These are the *trivial* divisors. A number is said to be *prime* if its only divisors are the trivial ones. The first primes are

$$2, 3, 5, 7, 11, 13, 17, 19, 23, \ldots.$$

Note that 1 is not a prime.

THEOREM. *Every natural number > 1 is a product of primes.*

The proof (induction by size) is left to the reader.

Our next theorem appears in Euclid's elements (300 BC).

THEOREM. *There are infinitely many primes.*

PROOF: Let $p_1, \ldots, p_n$ be n primes. Consider the number

$$p_1 \ldots p_n + 1.$$

By the preceding theorem, there is a prime p and a natural number a such that

$$ap = p_1 \ldots p_n + 1$$

and hence

$$ap - p_1 \ldots p_n = 1.$$

If p were one of the primes of the product, 1 would be divisible by p which is impossible. Hence, given any finite collection of primes, there is always a new one and this proves the theorem.

R. Every prime > 2 is odd and hence of the form $4k+1$ or $4k+3$ with k an integer. Prove that there are infinitely many primes of the form $4k+3$. (Hint. Let $q = p_1 \ldots p_n$ be a product of such primes. Factor $4q+3$.)

Remark. Dirichlet proved in 1837 that every arithmetic progression $ak + b$, where $k = 1, 2, \ldots$ contains infinitely many primes when a and b are natural numbers which have no non-trivial common factor.

Modules

Let Z be the set of integers,

$$Z = \{\ldots, -3, -2, -1, 0, 1, 2, 3, \ldots\}$$

The notion of divisibility carries over to all integers which are not zero with the difference that the trivial divisors of an integer a are $\pm a$ and ± 1. A non-empty subset M of Z is said to be a *module* or, more precisely, a Z-module, if

$$a, b \in M \;\Rightarrow\; xa + yb \in M$$

for all integers x and y. An example is the set Za of all integral multiples of a fixed integer a. Another example is the set $Za+Zb$ of all numbers of the form $xa + yb$ with x and y arbitrary integers.

R. Show that the definition of a module M amounts to M not being empty and

$$a, b \in M \;\Rightarrow\; a - b \in M.$$

(Hint. Show first that M contains 0 and with every a also $-a$.)

We shall now prove that *every* module M except zero has the form Zc for some natural number c. In fact, let c be the least positive number in M, and consider the module Zc contained in M. Marked out on the line, it appears as a grid of equidistant points as in the figure below. For the next few lines, the reader is referred to this figure.

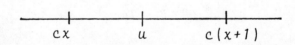

If u is any integer outside Zc, there is an integer x such that u is strictly between xc and $x(c+1)$. In particular, $u - xc$ is strictly between 0 and c. Hence, if u were also in M, M would contain $u - xc$ contradicting the definition of c. Hence $M = Zc$. This proves the first part of our next theorem, whose second part is left to the reader.

THEOREM. *To every pair of natural numbers a,b there is a unique natural number c such that*
$$Za + Zb = Zc.$$
The number c divides a and b and every number with this property divides c.

The number c is called the *greatest common divisor* (GCD) of a and b, and it is denoted by (a,b). We say that a is prime to b (and b prime to a) if $(a,b)=1$. The same situation is also described by saying that a and b are *coprime*.

R. Prove that if $(a,b)=1$, then $xa + yb = 1$ for some integers x and y (and conversely).

R. Prove that if p is prime and p divides a product ab, then p divides a or b. (Hint. Show that p divides b when $(p,a)=1$.)

Powers p^n, $n > 0$ an integer, of a prime p are the *primary* numbers. It follows from our last R that two primary numbers are coprime unless the corresponding primes are equal.

R. Prove the fundamental theorem of arithmetic: The factorization of a natural number > 1 into primes is unique apart from the order of the factors. (Hint. Any natural number is obviously a product of primary numbers belonging to different primes. Show that such a primary number is the largest power of the corresponding prime which divides the number in question.)

R. Show that a natural number a divides another one b if and only if every primary number dividing a also divides b.

Euclid's algorithm

Let $a > b > 0$ be natural numbers. To find their greatest common divisor, one employs Euclid's algorithm whose first step is
$$S : (a, b) \to (b, c)$$
where $r \geq 0$ and $< b$ is the remainder term when a is divided by b, $a = kb + r$. It is obvious that any number which divides a and b also divides b and r and conversely. In the second step of the algorithm, the pair (b, r) replaces

(a, b) and so on for the following steps. The algorithm ends when a pair $(c, 0)$ appears. The number c is then the greatest common divisor of a and b. In fact, by the preceding remark, every natural number which divides c also divides a and b and conversely.

Running the algorithm backwards produces in the end numbers x and y such that $ax + by = c$. In fact, each member of a pair in the algorithm is a linear combination with integer coefficients of the members of the preceding pair and a linear combination of linear combinations is still a linear combination.

Exercise

Show that if $a = 301, b = 211$, then Euclid's algorithm stops after 6 steps and that the two numbers are coprime. Also find numbers x and y such that $301x + 211y = 1$.

R. *Bases for the natural numbers.* Let $q > 1$ be a natural number. Using Euclid's algorithm and induction, prove that every natural number a has a unique expansion

$$a = a_0 + a_1 q + \cdots + a_n q^n$$

where $0 \leq a_j < q$ and n is the least number k with $a < q^{k+1}$.

R. Let $(b, c), (c, d), (d, e)$ be three consecutive pairs of Euclid's algorithm. Prove that $d < b/2$ and deduce from this that if an instance of the algorithm has $2n$ or $2n+1$ pairs and first element a, then $2^n < a$.

R. Let the pairs $(a_{k+1}, a_k), (a_k, a_{k-1})$ for $k = 0, \ldots, N-1$ be consecutive pairs of an instance of Euclid's algorithm. Prove that $a_k \geq b_k$ for all k where the b_k are the Fibonacci numbers defined recursively by $b_0 = 0, b_1 = 1$ and $b_{k+1} = b_k + b_{k-1}$ for $k = 2, 3, \ldots$. (Hint. Use that $a_{k+1} \geq a_k + a_{k-1}$.)

Exercises

1. Show that the equation

$$Zc = Zm_1 + \cdots + Zm_n$$

with integers $m_1, \ldots, m_n$ implies that c is their greatest common divisor.

2. Find integral solutions of $5x + 73y = 1$, of $112x + 6y = 2$ and of $112x + 6y = 4$.

3. Show that if the integers x, y satsify $ax + by = 1$, with integral a, b, c, then every other solution can be written $x + tb/d, y - td/b$ where $d = (a, b)$ and t is an arbitrary integer.

4. Show that $(a, c) = 1, (b, c) = 1$ implies $(ab, c) = 1$.

1.2 Congruences

Let m be a fixed integer. We say that x is *congruent* to y modulo m (or mod m for short) if m divides $x - y$. This is written as

$$x \equiv y \bmod m \text{ or } x \equiv y(m).$$

All numbers congruent to a given number x mod m form the *congruence class* $x + \mathbb{Z}m$ of x mod m. Computation mod m can be thought of as computation with these congruence classes as elements. A number in a congruence class is said to *represent* the class.

R. Show that there are precisely $|m|$ congruence classes mod m.

R. Show that $x \equiv y(m)$ and $z \equiv u(m)$ implies that $x \pm z \equiv y \pm u \ (m)$ and $xz \equiv yu(m)$.

Note. Let $C(x) = x + \mathbb{Z}x$ be the congruence class of x. The exercise above shows that the definitions

$$C(x) \pm C(y) = C(x \pm y), \quad C(xy) = C(x)C(y)$$

of sums, differences, and products of congruence classes are independent of the choice of representatives. It follows that $C(0)$ and $C(1)$ serve as zero and 1 for the congruence classes, and it is a simple but somewhat tedious matter to verify that the associative laws for addition and multiplication and also the distributive law hold for congruence classes precisely as for the integers (see section 3.1).

R. Given an integer a, show that there is another integer b such that

$$ab \equiv 1 \ (m)$$

if and only if a and m are coprime, $(a, m) = 1$.

A number b with the property above is called an *inverse* of a mod m and denoted by $a^{-1}(m)$. An inverse is unique mod m, for if $ab \equiv 1(m)$ and $ac \equiv 1(m)$, then $a(b - c) \equiv 0(m)$, so that, since $(a, m) = 1$, we have $b - c \equiv 0(m)$. Note that when m is prime, then every a not $\equiv 0(m)$ has an inverse mod m and that if $(a, m) = 1$, then the congruence $ax \equiv y(m)$ has the unique solution $x \equiv a^{-1}y(m)$. In this sense, the first congruence has been divided by a.

Simultaneous congruences

Consider simultaneous congruences

$$x \equiv a_1 \ (m_1), \ldots, x \equiv a_n \ (m_n)$$

where $m_1, \ldots, m_n$ are pairwise coprime, i.e., $(m_i, m_j) = 1$ unless $i = j$. These congruences can be solved step by step. In fact, according to the first one, $x \equiv a_1 + m_1 y$ for some y so that, substituting into the following ones,
$$m_1 y \equiv a_2 - a_1 \ (m_2), \ldots, m_1 y \equiv a_n - a_1 \ (m_n),$$
where we can divide by m_1 since this number is prime to m_2 etc. The result is $r - 1$ congruences of the form above with y as unknown. But we can also solve the congruences in one stroke by a method known as the

Chinese remainder theorem

Put $M = m_1 \ldots m_n$ and $M_k = M/m_k$. Clearly M_k and m_k are coprime for all k. Hence there are solutions c_k of the congruences
$$M_k c_k \equiv a_k \ (m_k)$$
for all k. The number
$$x = M_1 c_1 + \cdots + M_n c_n$$
solves the congruences above.

R. Verify this last statement. Show that the congruences are simultaneously solvable also when the left hand sides are replaced by $b_1 x, \ldots, b_n x$ and $(b_k, m_k) = 1$ for all k. Show that any solution is unique mod M.

Example

For $x \equiv 1 \ (2), x \equiv 2 \ (3), x \equiv 3 \ (5)$, the M_k are in order $15, 10, 6$ and the c_k are $1, 2, 3$ and we find that $x = 23$ satisfies all the congruences. The general solution is $53 + 30k$ with k arbitrary.

R. There is another way of representing the solution of the Chinese remainder problem above. Put
$$E_k = M_k b_k$$
with b_k chosen so that $E_k \equiv 1(m_k)$. Prove that $E_j E_k \equiv 0(M)$ when $j \neq k$ and $\equiv 1(M)$ otherwise. Prove that if a is given and $a \equiv a_k(m_k)$, then
$$a \equiv \sum E_k a_k \bmod M$$

R. Say that the a_k in this expansion are the coordinates of a. Prove that if b has the coordinates b_k, then $a_k \pm b_k$ and $a_k b_k$ are the coordinates of $a \pm b$ and ab, respectively.

Zeros of a polynomial modulo a prime

A polynomial
$$f(x) = a_0 + a_1 x + \cdots + a_n x^n$$
is said to be *unitary* or *monic* of degree n when $a_n = 1$.

THEOREM. *When p is a prime, and $f(x)$ is unitary with integral coefficients, the equation $f(x) \equiv 0(p)$ has at most n solutions mod p.*

Note the fact that $x^2 \equiv 1(8)$ has 4 solutions mod 8, namely $\pm 1, \pm 3$.

PROOF: The theorem is obvious for $n=1$. Let a be any integer. Writing $x = x - a + a$ and using the binomial theorem, we see that there is a unitary polynomial g with integral coefficients such that
$$f(x) = (x - a)g(x) + f(a)$$
If $f(a) \equiv 0(p)$, we have $f(x) \equiv (x - a)g(x)(p)$ for all integers x. Since p is a prime, $f(x)$ vanishes mod p if and only if $x - a$ or $g(x)$ or both vanish mod p. Hence an induction with respect to the degree finishes the proof.

Exercises
1. Discuss the congruence above when $(a, b) > 1$. Show that it has a solution if and only if (a, b) divides c.
2. Find all integers x for which i) $x \equiv 2(4)$, $x \equiv 4(5)$ ii) $x \equiv 1(2)$, $x \equiv 2(3)$, $x \equiv 3(5)$.
3. Let $f(x)$ be a polynomial with integral coefficients and let the integers m and n be coprime. Show that the number of roots mod mn of the equation $f(x) \equiv 0(mn)$ is the product of the number of roots mod m and mod n.

1.3 The theorems of Fermat, Euler, and Wilson

For a natural number m, let $\varphi(m)$ be the number of integers between 0 and m which are prime to m. The function φ is called *Euler's function*.

Example
$\varphi(2) = 1$, $\varphi(3) = 2$, $\varphi(6) = 2$, $\varphi(7) = 6$. When p is a prime, then $\varphi(p) = p - 1$.

THEOREM. (*Euler – Fermat*) *If a and m are coprime, then*
$$a^{\varphi(m)} \equiv 1 \ (m).$$

PROOF: Let $m_1, \ldots, m_k$ be the integers between 1 and m which are prime to m, so that $k = \varphi(m)$. Consider the numbers
$$am_1, \ldots, am_k.$$

They are all different mod m, for if $am_i \equiv am_j$, i.e. $a(m_i - m_j) \equiv 0(m)$, then m must divide $m_i - m_j$ since $(a,m) = 1$. Hence $m_i \equiv m_j(m)$, and the numbers above are the same mod m as the numbers $m_1, \ldots, m_k$. Multiplying we get

$$a^k m_1 \ldots m_k \equiv am_1 \ldots am_k \equiv m_1 \ldots m_k \ (m).$$

Here the product on the right is prime to m so that, dividing by it, we get the desired result.

COROLLARY. *(Fermat's Little Theorem)* If p is prime and a is not divisible by p, then
$$a^{p-1} \equiv 1 \ (p).$$

Note. Multiplying by a we get

$$a^p - a \equiv 0 \ (p),$$

a formula that holds for all a. It implies Fermat's theorem, for if a is not divisible by p, then it has an inverse b mod p. Multiplying by b, the result follows.

R. Prove our last formula in another way by noting that $(a+b)^p \equiv a^p + b^p$ mod p when p is a prime and using induction.

COROLLARY. *(Wilson's Theorem)* When p is a prime, then

$$(p-1)! \equiv -1 \ (p).$$

PROOF: Since the equation $x^{p-1} - 1 \equiv 0 \ (p)$ has the solutions $1, 2, \ldots, p-1$ and no others,

$$x^{p-1} - 1 \equiv (x-1)(x-2)\ldots(x-p+1) \ \text{mod} \ p$$

for all integers x. Putting $x = 0$ proves the corollary.

R. Prove the converse of Wilson's theorem.

Exercises

1. Verify that $3^{10} \equiv 1 \ (19)$ by explicit calculation.
2. Compute $\varphi(6), \varphi(32)$, and $\varphi(18)$ and verify that Euler's theorem holds for $m = 6$ and 32 and some $a > 1$.

Euler's function φ has some remarkable properties. We complete its definition by putting $\varphi(1) = 1$.

1.3 The theorems of Fermat, Euler, and Wilson

THEOREM. *When $q = p^k$ is primary and m and n are coprime, then*

$$\varphi(q) = q(1 - 1/p), \quad \varphi(mn) = \varphi(m)\varphi(n).$$

Note. It follows that

$$\varphi(m) = m \prod (1 - 1/p)$$

where m is any integer > 1 and p runs through the prime numbers dividing m.

Example
$\varphi(20) = 20 \cdot (1/2) \cdot (4/5) = 8.$

PROOF: The divisors of q are the integers $p, 2p, \ldots, q$ which are q/p in number. Hence

$$\varphi(q) = q - q/p = q(1 - 1/p)$$

which proves the first part of the theorem. To prove the second part consider integers of the form

$$z = mx + ny$$

where x and y are integers. If such a number is a multiple of mn, then $xm \equiv 0 \ (n)$ and $yn \equiv 0 \ (m)$ so that $x \equiv 0 \ (n)$ and $y \equiv 0 \ (m)$ since m and n are coprime. Next, let Z be the set integers z where x runs through the integers 1 to n and y through the integers 1 to m. Then Z has mn elements and, applying the result above to the differences of two integers of Z, we see that all elements of Z are incongruent mod mn. Hence they represent all congruence classes mod mn. Now an integer is prime to mn if and only if it is prime to both m and n. Hence an integer z in Z is prime to mn if and only if y is prime to m and x is prime to n. It follows immediately that $\varphi(mn) = \varphi(m)\varphi(n)$.

R. Show that $m = \sum \varphi(d)$ where d runs through all divisors of m. (Hint. Compute explicitly when m is a primary number and go from there.)

Moebius's Inversion Formula

Euler's function has an interesting connection with a function $\mu(n)$ from the natural numbers defined by Moebius. It is defined to be zero unless n is a product of k *separate* primes, in which case it equals $(-1)^k$. Hence it has the following properties

$$\mu(mn) = \mu(m)\mu(n), \ \mu(p^2) = 0, \ \mu(p) = 1, \ \mu(1) = 1$$

where m, n are coprime and p is prime.

R. Prove that $\sum \mu(d) = 0$ when d runs through the divisors > 1 of a number $n > 1$. (Hint. It suffices to consider the case when n is the product of k separate primes. It then reduces to the identity $(1-1)^k = 0$.)

THEOREM. *(Moebius's Inversion Formula)* Suppose that f and g are real functions defined on the positive divisors of a natural number n. Then the following two formulas, where d runs over the positive divisors of n, are equivalent,
$$f(n) = \sum g(d), \quad g(n) = \sum \mu(d) f(n/d).$$

Note. When $f = g = \varphi$ is Euler's function, we know the first formula to be true. The second one is the announced connection between φ and μ.

R. Prove the inversion formula by direct verification using the previous exercise. (Hint. That the second formula follows from the first is rather direct. To prove the first formula from the second, write the first formula as $f(n) = \sum g(n/d)$.)

Orders mod m

The set $\Pr(m)$ of integers prime to a given integer $m \neq 0$ has some interesting properties.

R. Let a in $\Pr(m)$ be given. Prove that all integers k such that $a^k \equiv 1\ (m)$ form a module.

The order of an integer a in $\Pr(m)$, denoted by $\operatorname{ord}(a)$, is defined as the least integer $n > 0$ for which $a^n \equiv 1(m)$. By the preceding exercise, all integers k such that $a^k \equiv 1(m)$ are integral multiples of $\operatorname{ord}(a)$.

R. Prove that the orders of $\Pr(15)$ are $1,2,4$ and that 2 has order $20 = \varphi(25)$ in $\Pr(25)$.

THEOREM. *When a and b are in $\Pr(m)$, the order of ab divides $\operatorname{ord}(a)\operatorname{ord}(b)$ and there is equality if and only if the orders of a and b are coprime. All the orders divide the maximal order of integers in $\Pr(m)$.*

PROOF: Let $j = \operatorname{ord}(a), k = \operatorname{ord}(b)$. Since $(ab)^{jk} \equiv 1(m)$, $\operatorname{ord}(ab)$ divides jk. If $(ab)^t \equiv 1(m)$, then $a^{tk} \equiv 1(m)$ and $b^{tj} \equiv 1(m)$ and hence t is a multiple of j and k. When j and k are coprime, this means that t is a multiple of jk so that $\operatorname{ord}(ab) = \operatorname{ord}(a)\operatorname{ord}(b)$. The last statement of the theorem follows if we can prove

LEMMA. *Let a,b be in $\Pr(m)$. Unless $\operatorname{ord}(b)$ divides $\operatorname{ord}(a)$, there is an element c in $\Pr(m)$ such that $\operatorname{ord}(c) > \operatorname{ord}(a)$.*

PROOF: Let $n = \operatorname{ord}(a)$, $k = \operatorname{ord}(b)$. Since k does not divide n, there is a prime p and a power $q \geq p$ of p such that pq divides k and q divides n but pq does not divide n. Put $n = qn'$, $k = pqk'$. Then the orders n' and pq of $a' = a^q$ and $b' = b^k$ are coprime. Hence the order of $c = a'b'$ is $n'pq >$ord(a). This finishes the proof.

Remark. The set CPr(m) of congruence classes of the elements of Pr(m) has $\varphi(m)$ elements and constitutes a commutative group under multiplication (see section 3.1). The properties proved above use only the group axioms.

In an appendix to this chapter we shall list all m for which Pr(m) is cyclic, i.e., consists mod m of all powers of a single element. An equivalent condition is that the maximal order of the elements of Pr(m) is $\varphi(m)$.

1.4 Squares and the quadratic reciprocity theorem

An integer a is said to be a square modulo another integer m if there is another integer b such that $b^2 \equiv a$ mod m. The number a is also said to be a *quadratic residue* mod m. Since the numbers 1 and 0 are their own squares, all integers are squares mod 2, but the study of squares modulo a prime $p > 2$ is a non-trivial matter. For $p > 2$ and $q = (p-1)/2$, we have

$$a^{p-1} - 1 = (a^q + 1)(a^q - 1).$$

Hence at least one of the factors is $\equiv 0(p)$.

THEOREM. *If a prime p does not divide a natural number a, then*

$$a^{(p-1)/2} \equiv 1 \text{ or } -1 \text{ mod } p$$

according as a is or is not a square mod p. There are as many squares as non-squares mod p.

PROOF: If a is a square, $a \equiv b^2(p)$, we have the first case by Fermat's theorem. If a is not a square, collect the numbers $1, \ldots p - 1$ in pairs x, x' such that $xx' \equiv a(p)$. Note that x and x' are different mod p since a is not a square. We get $q = (p-1)/2$ such pairs and hence

$$(p-1)! \equiv a^q \ (p)$$

so that the theorem follows from Wilson's theorem.

Example

It follows from the theorem that -1 is a square mod p if and only if $p \equiv 1(4)$.

For p a prime and $(a,p) = 1$, let the Legendre symbol $(a|p)$ (classical notation $(\frac{a}{p})$), denote 1 when a is a square mod p and -1 when a is not a square mod p. With this notation, the previous theorem says that $a^{(p-1)/2} \equiv (a|p)(p)$ when p is a prime and a is prime to p. The following important result will be proved in the third section of chapter 4.

THE QUADRATIC RECIPROCITY THEOREM. *Suppose that p and q are odd primes. Then*
$$(p|q)(q|p) = (-1)^{(p-1)(q-1)/4}.$$

Note. In other words, the product $(p|q)(q|p)$ equals 1 unless both p and q are $\equiv 3 \mod 4$, in which case it equals -1.

Note. This result, first proved by Gauss in 1801, is one of the most famous and beautiful results in number theory.

R. Define $(a|p)$ to be zero when p divides a. It is obvious that $(a|p) = (b|p)$ when $a \equiv b\ (p)$. Prove that $(ab|p) = (a|p)(b|p)$. (Hint: This amounts to the statement that the product of two squares mod p is a square mod p etc. At one point it is useful to know that there are as many squares as non-squares.)

R. Verify that $(-1|p) = (-1)^{(p-1)/2}$ and that $(a|2) = 1$ for all odd integers a.

Examples

Any odd prime has the form $6k + \epsilon$ where $\epsilon = \pm 1$. Hence $(p|3) = (6k + \epsilon|3) = (\epsilon|3) = \epsilon$, for 1 is a square mod 3 but not -1. Similarly, $p \equiv \pm 1$ or $\pm 2 \mod 5$. In the first case, $(p|5) = 1$, in the second $(p|5) = -1$.

Exercise

Do the same computations with 7 taking the place of 5.

The quadratic reciprocity theorem has the following complement:

THEOREM. *When $p > 2$ is a prime, $(2|p) = (-1)^c$, where $c = (p^2 - 1)/8$.*

Note. The proof below is similar to one of Gauss's proofs of the quadratic reciprocity theorem.

Note. Using this result and the quadratic reciprocity theorem, we can compute any $(n|p)$ with p prime. In fact, reducing n modulo p, we can assume that $1 \leq n < p$ and then factor n into powers of primes. Since $(ab|p) = (a|p)(b|p)$, this reduces the problem to the quadratic reciprocity theorem and the computation of $(2|p)$.

PROOF: We are going to consider the numbers $C = \{1, 2, \ldots, (p-1)/2\}$, whose sum is $c = (p^2 - 1)/8$, and the set $2C = \{2, 4, \ldots, p - 1\}$. Let

1.4 Squares and the quadratic reciprocity theorem

$A = \{a_1, \ldots, a_r\}$ be the numbers of $2C$ which are $> (p-1)/2$ and let $B = \{b_1, \ldots, b_t\}$ be the others. Then the numbers

$$D = \{p - a_1, \ldots, p - a_r, b_1, \ldots, b_t\}$$

are all > 0 and $\leq (p-1)/2$ and they form a permutation of the numbers in C. In fact, the first r ones are odd and different and the others are even and different. Now let $\pi(A)$ be the product of all the elements of A and the same for the other sets. By the construction of D, $(-1)^r \pi(D) \equiv \pi(A \cup B)$ mod p. On the other hand, $A \cup B = 2C$, and hence $\pi(A \cup B) = 2^{(p-1)/2} \pi(C)$. Hence

$$\pi(A)\pi(B) \equiv 2^{(p-1)/2}\pi(C) \equiv (-1)^r \pi(C) \bmod p.$$

By the first theorem of this section, $2^{(p-1)/2}$ equals $(2|p)$ mod p. Hence, since $\pi(C)$ is not divisible by p,

$$(2|p) = (-1)^r.$$

Next, let $|T|$ denote the sum of the elements of a set T. We have $|C| = c$ and

$$2|C| = |A| + |B|, \quad |C| = pr - |A| + |B|$$

so that, subtracting the two, $c = |C| = -pr + 2|A| \equiv -pr \equiv r \bmod 2$. It follows that $c \equiv r \bmod 2$ and this proves the theorem.

The Jacobi Symbol

Jacobi extended the Legendre symbol by putting

$$(a|N) = (a|p_1) \ldots (a|p_t)$$

when a is prime to N and N is the product of the primes $p_1, \ldots, p_t$, which need not be different. This symbol inherits from the Legendre symbol the multiplicative property

$$(ab|N) = (a|N)(b|N)$$

and if a is a square mod N, then a is a square modulo every prime p which divides N and hence $(a|N) = 1$. But $(a|N) = 1$ is no guarantee that a is a square mod N. We have, for instance, $(2|3) = (2|5) = -1$ so that $(2|15) = 1$, but $x^2 \equiv 2(15)$ implies, for instance, $x^2 \equiv 2(3)$.

The utility of the Jacobi symbol stems from its multiplicative property and the following partial extension of the quadratic reciprocity theorem.

THEOREM. *Let P and Q be odd, positive and coprime. Then*

$$(P|Q)(Q|P) = (-1)^{(P-1)(Q-1)/4}, \quad (2|P) = (-1)^{(P^2-1)/8}.$$

R. Deduce this theorem from the two preceding ones by showing that $\epsilon(P) \equiv (P-1)/2 \bmod 2$ has the property that $\epsilon(P_1 P_2) \equiv \epsilon(P_1) + \epsilon(P_2) \bmod 2$ when P_1 and P_2 are odd and that $\epsilon(P) \equiv (P^2 - 1)/8 \bmod 2$ has the same property. (Hint. Write odd numbers as $4k+1$ or $4j+3$.)

Note. By this theorem and the property of the Jacobi symbol that $(P|Q) = (R|Q)$ when $P \equiv R \bmod Q$, any Jacobi or Legendre symbol can be computed without factoring into primes. Example: since 25 has the form $4k+1$, $(25|63) = (63|25) = (13|25) = (25|13) = (12|13) = (2|13)^2(3|13) = (13|3) = (1|3) = 1$. In the next chapter the Jacobi symbol will appear in a primality test.

Exercises
1. Show that $(3|73) = 1$ and that $(17|73) = -1$.
2. Show that 2 is a quadratic residue of every prime of the form $8n \pm 1$ but not a quadratic residue of the primes of the form $8n \pm 3$.
3. Show that there are infinitely many primes of the form $4k+1$. (Hint. Let $p_1, \ldots, p_n$ be such primes and put $N = 4(p_1 \ldots p_n)^2 + 1$. If p is a prime dividing N, then -1 is a square mod p.)

1.5 The Gaussian integers

Number theory has many objects which are called integers without being the usual or *rational* integers. Complex numbers of the form $a + ib$ where a and b are rational integers are called *Gaussian* integers. The set of these numbers is denoted by $Z[i]$. It is clear that sums, products and the negatives of Gaussian integers are Gaussian integers. Divisibility is defined in the natural way, i.e. x divides y if there is a Gaussian integer z such that $y = xz$. The concept of a prime number can be transferred to the Gaussian integers, but not without surprises. There are rational primes, i.e. primes in Z, which are not primes in $Z[i]$, e.g., $5 = (2+i)(2-i)$ and $2 = (1+i)(1-i)$. When $x = a + ib$ is a Gaussian integer, let $N(x)$ be its absolute value squared,

$$N(x) = |x|^2 = a^2 + b^2.$$

It is clear that $N(xy) = N(x)N(y)$. If $N(x) = 1$, x is called a *unit*. Units are not considered to be primes. The following exercises are easy transplants of the corresponding ones for the integers.

R. Show that $1, -1, i, -i$ are the only units and that every Gaussian integer is a unit times a product of primes.

R. Show that the Gaussian integers have an infinity of primes. (Hint: copy the proof for rational integers.)

As we did for the rational integers, we can consider Z[i]-modules or Gaussian modules of Z[i], defined as non-empty subsets M of Z[i] such that

$$x, y \in M \Rightarrow ux + vy \in M$$

for all Gaussian integers u and v (it suffices that M is not empty and that $x, y \in M \Rightarrow x - y \in M$ and $ix \in M$). All Gaussian integral multiples of a fixed Gaussian integer z is such a module. We note that it consists of the numbers $0, z, iz, z + iz$ which are the corners of a square Q in the complex plane. Geometrically, M consists of all corners of all squares $Q + nz + imz$ with m and n in Z forming a square net in the complex plane. See the figure below.

R. Prove the last statement.

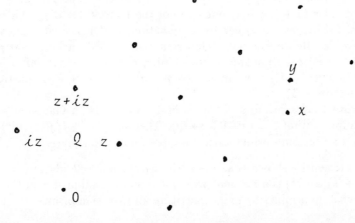

Figure. The dots denote complex numbers of the form $Q+nz+imx$ where m, n are rational integers and Q a square with the corners $0, z, iz, z + iz$ with z a fixed Gaussian integer.

Now suppose that M is a Gaussian module and let z be a member of M for which $N(z)$ is minimal among all the non-zero numbers in M. In other words, the distance to 0 among all the other points of M is minimal for z. Next, let y be in M. Then y belongs to some square of the net corresponding to Z[i]z so that there must be a point x of this net whose distance to y is at most $|z|/\sqrt{2}$, i.e., $N(x - y)$ is at most $N(x)/2$. Since $x - y$ is in M, this contradicts the property of z unless $x = y$. Hence

THEOREM. *Any Gaussian module consists of all Gaussian multiples of a fixed Gaussian integer.*

We can now repeat the argument used for the integers to obtain a divisibility theory for the Gaussian integers.

R. Formulate and prove an analogue for the Gaussian integers of the fundamental theorem of arithmetic.

In addition, we can prove a remarkable result of number theory proved by Fermat by another method.

THEOREM. *Every rational prime of the form $4k+1$ is the sum of the squares of two integers.*

R. Show that an odd number which is a sum of the squares of two integers must have the form $4k+1$.

PROOF: Let p be such a prime. Then, as we have seen, -1 is a square mod p. In other words, p divides q^2+1 for at least one integer q. But then p divides $(q+i)(q-i)$ but none of the factors, for if p divides one of them, it divides the other by conjugation, and hence also $2i$, which is impossible. Hence p is not a Gaussian prime. We conclude that p is a product $(a+ib)(c+id)$ of two Gaussian integers which are not units. Hence $p^2 = (a^2+b^2)(c^2+d^2)$ so we must have $p = a^2+b^2 = c^2+d^2$, in particular, $p = (a+ib)(a-ib)$.

R. Prove that a prime p of the form $4k+3$ is also a Gaussian prime. (Hint. Suppose that $p = (a+ib)(c+id)$. Then $p^2 = (a^2+b^2)(c^2+d^2)$ and, if p is not a Gaussian prime, then p is the sum of two squares.)

Note. It can be shown that $1+i$, the numbers $a+ib$ above (i.e., such that $(a+ib)(a-ib)$ is a rational prime) and all rational primes of the form $4k+3$ when multiplied by units constitute all Gaussian primes.

1.6 Algebraic numbers

A real or complex number a is said to be *algebraic* if it is a root of some equation

(1) $$x^n + c_{n-1}x^{n-1} + \cdots + c_0 = 0$$

with *rational* coefficients c_k. The least number n for which this situation holds is called the *degree* of a. When all the coefficients above are rational integers, a is said to be an *algebraic* integer. (Note that the polynomial above is unitary.)

1.6 Algebraic numbers

Using the language of linear algebra, we can say that a is algebraic of degree at most n if the power a^n of a is a linear combination with rational coefficents of its powers with exponent $< n$.

R. Show that in this case, *all* powers of a are linear rational combinations of the powers with exponent $< n$.

We can rephrase this last result as follows where Q denotes the rational numbers: If a is algebraic of degree n, then all powers of a lie in the set

$$Q + Qa + \cdots + Qa^{n-1}$$

The following fundamental lemma shows that this formulation is well chosen.

LEMMA. *Suppose that $x_1, \ldots, x_n$ are complex numbers and that x^m belongs to*

$$Qx_1 + \cdots + Qx_n$$

for all $m \geq 0$. Then x is algebraic.

PROOF: By assumption, the numbers $1, x, \ldots, x^n$ are linear combinations with coefficients in Q of $x_1, \ldots, x_n$. Hence, if $y_0, \ldots, y_n$ are any numbers, then

$$y_0 + y_1 x + \cdots + y_n x^n = f_1(y) x_1 + \cdots + f_n(y) x_n,$$

where $f_1, \ldots, f_n$ are linear combinations of $y_0, \ldots y_n$ with coefficients in Q. Now the system of equations for $y_0, \ldots y_n$,

$$f_1(y) = 0, \ldots, f_n(y) = 0$$

is linear and it has n equations with $n + 1$ unknowns. Hence, by linear algebra, it has a non-trivial solution

$$y_0 = a_0, \ldots, y_n = a_n$$

with rational right sides. Hence x is algebraic.

Suppose now that all powers of x and y are rational linear combinations of all products $x_j y_k$. Then, if a and b are rational, the binomial theorem shows that all powers of $ax + by$ are rational linear combinations of all products $x_j y_k$. This proves an important result.

THEOREM. *Rational multiples, sums and products of algebraic numbers are algebraic.*

R. If x is algebraic and not 0, prove that $1/x$ is algebraic.

R. Using the proof of the theorem above, show that if x and y are algebraic of degrees m and n respectively, then $x+y$ and xy are algebraic of degree at most mn.

Algebraic integers

Next, consider the situation when $x_1,\ldots,x_n$ are complex numbers such that one of them is 1 and all products $x_j x_k$ are linear combinations with *integral* coefficients of $x_1,\ldots,x_n$.

LEMMA. *Under these conditions, every x in $Zx_1+\cdots+Zx_n$ is an algebraic integer.*

PROOF: We have
$$xx_j = a_{j1}x_1 + \cdots + a_{jn}x_n$$
with integer a_{jk}. This can be written as
$$\begin{aligned}(a_{11}-x)x_1 + &\quad \cdots &\quad \cdots + a_{1n}x_n &= 0,\\ a_{21}x_1 + &\quad (a_{22}-x)x_2 + &\quad \cdots + a_{2n}x_n &= 0,\\ &\cdots\\ a_{n1}x_n + &\quad \cdots + &\quad (a_{nn}-x)x_n &= 0.\end{aligned}$$
If A is the matrix (a_{jk}), and $D(x) = \det(A - xI)$, I the unit matrix, this gives $D(x)x_k = 0$ for all k so that $D(x) = 0$. Since $D(x) = (-1)^n x^n + \ldots$ has integral coefficients, x is entire algebraic.

R. Prove that integer multiples, sums and products of algebraic integers are algebraic integers.

Exercises

1. We know that $\sqrt{2}+\sqrt{3}$ is an algebraic integer, i.e. the root of a unitary polynomial with integral coefficients. Find such a equation and all its roots.
2. Show that $2\cos 2\pi/n$ is an algebraic integer for every integer $n > 0$. (Hint. Write it as the sum of two algebraic integers.)
3. Let x be an algebraic number. Show that mx is an algebraic integer for some natural number m.

1.7 Appendix. Primitive elements and a theorem by Gauss

When $\varphi(m)$ is the maximal order of $\Pr(m)$ or, in other words, when $\Pr(m)$ has an element a of order $\varphi(m)$, we say that $\Pr(m)$ is *cyclic* and that a is a primitive element or a generator of $\Pr(m)$. Our next aim is the existence and construction of such elements. The proofs give a taste of serious number theory with plenty of separation of cases.

1.7 Appendix. Primitive elements and a theorem by Gauss

THEOREM. *When $m = p$ is prime, the maximal order is $p - 1$.*

PROOF: Let k be the maximal order. Then, since every order divides the maximal order, $x^k \equiv 1(p)$ for all numbers in Pr(p). But there are at most k integral solutions mod p of this equation (see the last theorem of section 1.2). Hence $k = p - 1$.

To continue we shall use the following lemma.

LEMMA. *When $p > 2$ is a prime, $(u, p) = 1$, $f > 0$ and $r > 0$ are integers, then*

(1) $$(1 + up^f)^{p^r} \equiv 1 + up^{f+r} \mod p^{f+r+1}$$

The same holds for $p = 2$ and $f > 1$ while

(2) $$(1 + 2u)^{2^r} \equiv 1 \mod 2^{r+2}$$

PROOF: When $p > 2$, the binomial coefficients $\binom{p}{k}$ are divisible by p unless $k = 0$ or p and this proves (1) when $r = 1$. Taking the pth power proves the formula for $r = 2$ and so on. If (1) holds for $p = 2$ and $f > 1$, then, squaring the right side gives a number

$$\equiv 1 + u2^{f+r+1} + u^2 2^{2f+2r} \mod 2^{f+r+2}.$$

Since $r > 0, f > 0$, we have $2f + 2r \geq f + r + 2$ so that (1) holds for $r + 1$. Finally, for $f = 1$ we have

$$(1 + 2u)^2 = 1 + 4u + 4u^2 = 1 + 8v$$

where v is an integer. Induction in the manner above then proves that (2) holds. The proof is finished.

THEOREM. *Let $q = p^f$ be primary. When $p > 2$, the maximal order mod q is $\varphi(q) = (p-1)p^{f-1}$ and the primitive elements a are the primitive elements mod p for which $a^{p-1} \not\equiv 1 \mod p^2$. When $p = 2$, the maximal order mod q is $\varphi(q)/2 = 2^{f-2}$, and the elements of maximal order are the numbers $\equiv 3$ (4) when $q = 4$ and the numbers $\equiv 3, 5$ (q) when $q \geq 8$.*

PROOF: Let $p > 2$ and let a have the properties listed. Then

(3) $$a^{p-1} \equiv 1 + up \mod p^2, \quad (u, p) = 1.$$

Hence the order of a is $\varphi(q)$ by the formula (1). The same formula shows that a cannot be a primitive element when $a^{p-1} \equiv 1 \mod p^2$. To complete the first part of the theorem, we have to prove that there are elements a

with the property (3). This is clear. In fact, take a b such that $b^{p-1} \equiv 1$ mod p^2 and put $a = b + p$. Then

$$a^{p-1} \equiv 1 + b^{p-2}(p-1)p \text{ mod } p^2$$

so that a has the property (3). This proves the first part of the theorem.

The second part is a matter of immediate verification except when $f > 2$ in which case (1) and (2) show 3 and 5 to have the order $2f - 2$ mod 2^f. The larger odd numbers > 2 have the form $1 + 2^t u$, u odd and $t > 2$ and the formulas (1) and (2) prove that they are not primitive elements mod 2^f. This finishes the proof.

We can now prove a general result due to Gauss.

THEOREM. *The maximal order of* $\Pr(m)$ *divides the least common multiple of the numbers* $\varphi(q)$ *where q runs through the primary divisors of m. This order equals* $\varphi(m)$ *if and only if* $m = 2, 4, p^k, 2p^k$ *where p is an odd prime.*

PROOF: Let m_k for $k = 1, \ldots, n$ be the primary factors of m. Let M be the least common multiple of $\varphi(m_1), \ldots, \varphi(m_n)$. By Euler's theorem, $a^M \equiv 1(m_i)$ for every i and hence $a^M \equiv 1(m)$. It follows that the maximal order of $\Pr(m)$ is at most the least common multiple of the numbers $\varphi(m_i)$ and hence less than $\varphi(m)$ when two of the numbers $\varphi(m_i)$ have a common factor > 1. Since $\varphi(q) = (q/p)(p-1)$ is divisible by 2 when q is a power of an odd prime or q is a power of 2 greater than 4, this excludes every m with two primary factors belonging to odd primes. It also excludes every m which is the product of a primary factor and a power of 2 greater than 2. The remaining cases for m are those listed in the theorem and, since $\varphi(2) = 1$, $\varphi(4) = 3$, also the powers of 2 greater than 4. But they are excluded by the previous theorem.

It remains to exhibit primitive elements for the cases listed. When $m = p^k, k > 0, p$ a prime $> 2, \Pr(m)$ is cyclic and the previous theorem gives the primitive elements. The cases $m = 2, 4$ are true by direct verification. Finally, let $m = 2p^{k+1} = 2q$ and let a have order $\varphi(q)$ mod q. Let b be one of the numbers a and $a + q$, whichever one is odd. Then b has order $\varphi(2q)$ mod $2q$. This finishes the proof.

Literature

The first systematic account of number theory is *Disquisitiones Arithmeticae* by Gauss, first publishes in 1801 (also Springer 1986) and thereafter a model for all books on elementary number theory. The present chapter covers about the first quarter of *Disquisitiones*. The Gaussian integers appeared for the first time in an article from 1832.

CHAPTER 2

Number theory and computing

Number theory has many connections with computer science. Some of them are touched upon in this chapter, namely the cost of arithmetic operations and the use of the Chinese remainder theorem, the cost of identifying primes and factoring into primes and the Public key code and, finally, a classical construction of pseudo-random numbers in which number theory plays a part.

2.1 The cost of arithmetic operations

One of the branches of computer science has to do with estimates of running time or cost of arithmetic operations. In order to get very precise results, one has to take into account the capabilities of the machine used. Here we shall only derive rough upper bounds which have some credibility for the handling of large numbers.

Addition and subtraction

In practice, numbers are written in systems whose base is a power of 2, mostly 16. We shall consider the simplest case when the base is 2 and we shall write

$$N = [\ldots abc.def \ldots]$$

when

$$N = \ldots 4a + 2b + c + 2^{-1}d + 2^{-2}e \ldots$$

where each coefficient is 0 or 1. An integer whose format is n binary digits will be called an n-bit integer.

When a computer adds two n-bit integers, it proceeds in a number of steps realized by one action of a logical gate. Each step, for instance adding 1 and 1, is counted as a *bit operation*. Leaving the details of this aside, we are going to start from the axiom that addition or subtraction of two n-bit integers requires at most a fixed constant times n bit operations, or, for simplicity, $O(n)$ bits. Without specification of the constant, such an estimate is useless in practice, but the estimates below are of some interest for the handling of large numbers.

Multiplication

The classical algorithm for multiplying two n-bit integers amounts to adding n suitably shifted n-bit integers. If we perform the additions successively, we may eventually have to deal with close to $2n$-bit integers and hence multplication of two n-bit integers costs at most $O(n^2)$ bits. We get the same estimate if we write our integers as

$$u = \sum 2^j u(j), \quad v = \sum 2^k v(k),$$

where the sums run from 0 to n. Then

$$uv = \sum 2^r w(r), \quad w(r) = \sum u(j)v(k)$$

where r runs from 0 to $2n$ and $j+k = r, 0 \leq j, k \leq n$ in the second sum. The number of terms in the sum for $w(r)$ is at most $n-1$ and this happens when $r = n$. Hence w costs $O(n^2)$ bits to compute with no immediate prospect for improvement. The following trick reduces our present estimate somewhat.

Let us write two $2n$-bit integers if the form $a = x + 2^n y, b = z + 2^n u$ where x, y, z, u are n-bit integers. The product

$$ab = xz + 2^n(xu + yz) + 2^{2n}uy$$

apparently requires 4 multiplications of at most $2n$-bit integers. But if we write

$$xu + zy = (x + z)(u + v) - xz - uy.$$

we need only 3 multiplications (and 4 additions) of at most $2n$-bit integers. The idea is now to make an induction from n to $2n$ to estimate the total bit cost under repetitions of the trick. Let $c(n)$ be the bit cost of multiplying in this way the product of two n-bit integers. Our step requires 3 multiplications of n-bit integers and 4 additions of at most $(n + 1)$-bit integers. The cost of the additions is at most $2An$ for some $A > 0$. Hence

$$c(2n) \leq 3c(n) + 2An.$$

Since this recursion formula implies that

$$c(2n) + 2A2n \leq 3(c(n) + 2An)$$

when $n \geq 1$, we get

$$c(2^k) = O(3^k) = O(2^{ck})$$

where $c = \log_2 3 = 1.76\ldots$. This improves the previous result when n is a power of 2. Enclosing n between to successive powers of 2 proves that $c(n) = O(n^c)$.

Remark. In section 4.2, the fast Fourier transfrom will be used to compute the coefficients $w(r)$ above. This together with a choice of basis depending on n gives another method, the Schönhage-Strassen algorithm. It multiplies two n-bit integers at the bit cost of $O(n \log n \log \log n)$.

Division

There is high precision algorithm for computing the inverse of a number $A > 0$ which depends on the properties of the function $f(x) = 2x - Ax^2$. It is positive when $0 < x < 2/A$ and has a maximum $1/A$ when $x = 1/A$. Hence an iteration

$$x \to f(x)$$

should increase rapidly to $1/A$ if we start in the interval between 0 and $1/A$.

Reciprocal

Let

$$A = [1.a(1)a(2)\ldots]$$

be a real number and let $A(j) = [1.a(1)\ldots a(2^j)]$. Approximations $B(k)$ to $1/A$ are given by the recursion formula

$$B(0) = [0.1], \quad B(k+1) = 2B(k) - A(k+1)B(k)^2$$

where $k = 0, 1, 2, \ldots$.

That the algorithm has quadratic convergence follows from

LEMMA. *The approximations $B(k)$ to $1/A$ satisfy the following inequalities*

$$0 \le B - B(k) \le 2^{-2^k}$$

for all k.

PROOF: It suffices to prove that $S(k) = 1 - A(k)B(k)$ satisfies the same inequality as $B - B(k)$. In fact, we have $AB = 1$ and hence, if $A = A(k) + A'(k)$, $B = B(k) + B'(k)$, then

$$A(k)B'(k) + A'(k)B(k) + A'(k)B'(k) = 1 - A(k)B(k)$$

where all the terms on the left are positive. Hence $B'(k) \le S(k)$ with strict inequality when $A(k) > 1$.

Clearly $A(0)B(0) = [0.11]$ when $a(1) = 1$ and $A(0)B(0) = [0.1]$ when $a(1) = 0$ so that the desired inequality for $S(k)$ holds for $k = 0$. Now, if $A(k+1) = A(k) + C(k)$, the recursion formula shows that $S(k+1) = 1 - A(k+1)B(k+1)$ equals

$$1 - 2B(A+C) + B^2(A+C)^2 = (1 - BA - BC)^2 = (S - BC)^2,$$

where A, B, C, S have the argument k. By induction, $0 \leq S(k) \leq 2^{-2^k}$ and by construction, $0 \leq C(k) \leq 2^{-2^k}$. Also,

$$B(k) \leq A(k)B(k) = 1 - S(k) < 1.$$

Hence $S(k+1) \leq 2^{-2^{k+1}}$ which finishes the proof.

Although our algorithm computes n digits of the reciprocal of an n-bit integer in about $\log n$ steps, the step k requires three multiplications and one addition of k-bit integers. Hence the cost of the step k is at most

$$3M(2^{k+1}) + A2^{k+1}$$

for some A. Here $M(2^k)$ is the bit cost of computing the product of two k-bit integers. If we make the reasonable assumption that

(1) $$M(n/2) \leq cM(n)$$

for n a power of 2 and some $c < 1$, we can sum the total bit cost of our algorithm, say $R(n)$, to at most

$$O(M(n) + n) = O(M(n))$$

when n is a power of 2. It is also possible to go the other way. Since $4ab = (a+b)^2 - (a-b)^2$, we have $M(n) = O(S(n))$ where $S(n)$ is the bit cost of squaring two n-bit integers. One also has

$$P(P+1) = \frac{1}{\frac{1}{P+1} - \frac{1}{P}},$$

from which it follows that $M(n) = O(R(n))$. Hence we have proved

THEOREM. *Under the hypothesis (1), the bit costs of multiplying, squaring and inverting n-bit integers are of the same order.*

Remark. One simple, interesting and useful way of looking at the cost of arithmetic without division is to assume that every multiplication costs one unit and addition and subtraction nothing. Consider for instance the cost of computing the value of a polynomial

$$f(x) = a_0 + a_1 x + \cdots + a_n x^n.$$

As it is written, the cost is $2 + 3 + \cdots + n + 1 = n(n+3)/2$. A better way is to use Horner's rule: compute in order

$$a_n x, (a_n x + a_{n-1} x) x, \ldots$$

which gives the cost n. That this is best possible will be shown in Chapter 6 as an application of the theory of polynomial rings.

Arithmetic using the Chinese Remainder Theorem

Suppose that the positive integers

$$m_1, \ldots, m_n$$

are pairwise coprime and let M be their product. According to one version of the Chinese remainder theorem (see section 1.2), every integer is congruent mod M to a sum

$$u = u_1 E_1 + \cdots + u_n E_n$$

where the coefficients u_k of u are residues of u mod m_k and the numbers E_k have the property that $E_k \equiv 1 \mod m_k$ and $E_k^2 \equiv 0 \mod M$ when $j \neq k$. The integers $E_k \equiv t_k M_k$ are constructed by finding integers t_k such that

$$t_k M_k \equiv 1(m_k), \quad M_k = M/m_k.$$

These formulas reduce addition, subtraction and multiplication mod M to the corresponding operations mod m_k for all k. In fact, the coefficients of $u \pm v$ and uv are given by the residues of these numbers mod m_k.

The gain may be illusory since the computation of the numbers t_k and the M_k and E_k involves multiplication of large numbers. But when the E_k are computed once and for all and many computations have to be made there might be some gain over multiplication mod M. The numbers t_k are obtained using Euclid's algorithm. The computation of the M_k may be facilitated by successive partitions into groups. For instance, when $n = 4$, then $m_{12} = m_1 m_2$ and $m_{34} = m_3 m_4$ are computed first.

Note. If $M(b)$ majorizes the bit cost of computing the product of two b-bit numbers, the total bit cost of computing the numbers E_k when the M_k are b-bit numbers has been estimated to $O(M(bn) \log n + O(nM(b) \log b))$ (see Aho-Hopcroft-Ullman (1974)).

2.2 Primes and factoring

Most of the properties of primes can be used to prove that an integer is composite. For instance

THEOREM. *A natural number N is a prime if and only if for every prime p dividing $N - 1$, there is an integer a such that*

$$a^{N-1} \equiv 1(N), \quad a^{(N-1)/p} \not\equiv 1(N).$$

PROOF: Let $\Pr(N)$ be the set of integers prime to N. When N is a prime, just choose the order of a to be $\varphi(N) = N - 1$. Otherwise, let n be the order of a in $\Pr(N)$. By the first condition, n divides $N - 1$ and by the the second condition, $(N - 1)/p$ is not a multiple of n. Hence the primary decompositions of N and n must have the same power q of p. It follows that $\varphi(N)$, of which n is a factor, has the factor q. Hence $\varphi(N) \leq N - 1$ has all the primary factors of $N - 1$ so that $\varphi(N - 1) = N - 1$ and N must be a prime.

This criterion, which is due to Lehmer, is especially efficient when applied to Fermat numbers

$$F(n) = 2^{2^n} + 1$$

among which there are primes and non-primes. Lehmer's criterion contains the sufficiency part of the following result.

PEPIN'S THEOREM. *A necessary and sufficient condition for $F(n)$ to be a prime is that*

$$3^{(F(n)-1)/2} \equiv -1 \mod F(n).$$

The necessity is a consequence of the quadratic reciprocity theorem. In fact, $F(n) \equiv -1(3)$ so that $(F(n)|3) = -1$. In addition, $(F(n)-1)(3-1)/4$ is even and hence $(3|F(n)) = -1$ when $F(n)$ is a prime and this implies the equality above.

The properties of the Jacobi symbol gives another primality test. Let $J(N)$ be the set of congruence classes mod N which are prime to N and satisfy the congruence

(1) $$(a|N) \equiv a^{(N-1)/2} \mod N.$$

THEOREM. *When N is odd and not a prime, $J(N)$ is a proper subgroup of $\Pr(N)$.*

Note. If N is a prime, $J(N) = \Pr(N)$, (section 1.4). If $J(N)$ is a proper subgroup of $\Pr(N)$, its order is at most half that of $\Pr(N)$. Solovay and Strassen (1977) have the same result with an incomplete proof.

PROOF: Since $\Pr(N)$ is a group and the Jacobi symbol and the right side of (1) are multiplicative, $J(N)$ is a group. Suppose that $J(N) = \Pr(N)$ so that (1) holds for all a in $\Pr(N)$. If $N = p^k$ is primary but not prime, $\Pr(N)$ is cyclic of order $\varphi(N) = p^{k-1}(p - 1)$ by Gauss' theorem. Squaring

(1), this order divides $N-1$ which is impossible. Hence $N = rs$ with r,s coprime. If there is an a in $\Pr(N)$ with $(a|N) = -1$, using the Chinese remainder theorem, we can choose b in $\Pr(N)$ with $b \equiv a \mod r$ and $b \equiv 1 \mod s$. Then $b \equiv a \mod N$

$$b^{(N-1)/2} \equiv (a|N) = -1 \mod r, \quad b^{(N-1)/2} \equiv 1 \mod s.$$

Hence, if (1) holds for b, it follows that $(b|N) = 1$ and -1 at the same time which is impossible. Hence, if $J(N) = \Pr(N)$, then $(a|N) = 1$ for all a in $\Pr(N)$. Let

$$N = \Pi p^{m(p)}$$

be the primary decomposition of N. Let q be one of the primes in the product and choose a number c such that $(c|q) = -1$. By the Chinese remainder theorem, there is an a such that $a \equiv c \mod q$ and $a \equiv 1 \mod p$ when $p \neq q$. But then a and N are coprime and, by the definition of the Jacobi symbol, $(a|N) = (-1)^{m(p)}$. It follows that N has to be a square when $(a|N) = 1$ for all a in $\Pr(N)$. But if $N = M^2$ with an an integral M, $1 + M$ is prime to N and putting $a = 1 + M$ in (1), squaring and using the binomial theorem gives $-M \equiv 0 \mod M^2$ which is a contradiction. The proof is finished.

Some primality tests are based on trial and error and have the property that a given number is composite decreases considerably with every step of the algorithm. We shall describe one such test due to Solovay and Strassen (1977), which uses the preceding theorem. It has one repetitive step with the following substeps:

1) Choose a at random between 2 and N.
2) if $(a, N) > 1$ end.
3) if $(a, N) = 1$ and (1) does not hold, end.
4) go to 1).

By the theorem and the note after it, the chance of N being composite when the algorithm has not stopped after one step is at most $1/2$. Hence, if the algorithm has not stopped after n steps, the chance that N is composite is at most 2^{-n}.

The practical value of the test is of course bound by the cost of computing (a, N) and the two sides of (1). Since Euclid's algorithm for the pair N, a has at most $O(\log N)$ steps, (see the exercises p. 4), the bit cost of computing (a, N) is also $O(\log N)$. The cost of computing the right side of (1) has the same upper bound (Knuth (1977) p. 409) and the generalized quadratic reciprocity formula (p. 14) makes the cost of computing the Jacobi symbol $(a|N)$ comparable to that of (a, N) and hence it is at most $O(\log N)$. Hence this is also the cost of carrying through one step of the algorithm above.

Factoring large numbers

The methods of factoring large numbers (only odd ones are relevant) are all more or less based on trial and error. A simple first step is to have a list of the first primes and use Euclid's algorithm to check whether they are factors. If the given number is N, it is of course only necessary to check the primes $\leq \sqrt{N}$. If the primes available are too few to reach this limit or if one wants to proceed faster, other methods are necessary. Most of them are listed in Knuth (1981) and some are treated in Riesel (1985). Here we shall only mention one method which has been very successful (see Knuth for references). It uses a given collection of primes in a clever way.

The first step is to look for integers x, y between 0 and N such that $x^2 - y^2 \equiv 0(N)$ but $x + y \not\equiv 0(N)$. Then N has the proper factor $x - y$. The second step is to look for squares mod N which are close to N and can be factored mod N into products

$$x^2 \equiv (-1)^{e(0)} p_1^{e(1)} \ldots p_n^{e(n)} \mod N$$

of primes in a given collection $p_1, \ldots p_n$. If a set $\{x_1, \ldots, x_r\}$ of such numbers x have been found with the property that the sum of the vectors of their exponents has even components $2f(0), \ldots, 2f(n)$, then

$$x \equiv x_1 \ldots x_r, \quad y \equiv (-1)^{f(0)} p_1^{f(1)} \ldots p_n^{f(n)}$$

have the property that $x^2 - y^2 \equiv 0(N)$ and except for the mishap that $x \pm y \equiv 0(N)$, a proper factor of N has been found.

This method was used to prove that the Fermat number $2^{128} + 1$ is not a prime. Its bit cost has been estimated to $N^{\epsilon(N)}$ where $\epsilon(N) = O(\sqrt{\log N \log \log N})$.

Trapdoors and Public Key

A *trapdoor* function is a bijection of a set M such that its values $f(x)$ are easy to compute but the inverse of f is difficult to compute without some secret information. The best known instance is the Public Key or RSA code after Rivest, Shamir and Adleman (1978). It relies on the following piece of number theory.

THEOREM. *If N is a product of distinct primes p and $t(N)$ is the least common multiple of all $\varphi(p)$, then*

$$a^{t(N)+1} \equiv a \mod N$$

for all integers a.

PROOF: It suffices to prove that the desired congruence holds modulo all prime factors p of N. When $a \equiv 0(p)$, the congruence holds modulo p. When $(a,p) = 1$, then $a^{p-1} \equiv 1(p)$ by Fermat's theorem and hence also $a^{t(N)} \equiv 1(p)$ since $t(N)$ is divisible by $p-1$. Hence $a^{t(N)+1} \equiv a(N)$ always, and this proves the theorem.

This theorem and the difficulties of factoring large numbers if the basis of the *Public key*. Suppose that N is the product of distinct and very large primes and suppose that there are two positive integers k and k' such that $kk' = t(N) + 1$. Then any message which can be represented by a number $a > 0$ and $< N$ can be coded as $b = a^k$ and recovered by raising b to the power k' and doing the computations modulo N. A possible use of this is the following. A person A wants to receive messages from a circle C of other persons but he wants the messages to be kept secret during the transmission. He then arranges for the persons in C to know the numbers k and N and the encoding procedure $a \to a^k$. Then A is not in trouble if the encoding procedure becomes public knowledge. In fact, decoding the message requires knowledge of the number k' which can more easily be kept secret. One way of breaking the code is to guess the coding principle and then factor N which requires a computer and the knowledge of a specialist. Hence decoding seems practically impossible and this is the origin of the term Public Key.

Note. It has been guessed but not proved that breaking the Public Key is equivalent to factoring.

2.3 Pseudo-random numbers

Sometimes a computer is required to produce something like random numbers in an interval. The word random taken in the strict sense means that every number has the same probability of occurring regardless of the preceding ones. This ideal situation is not possible in practice. A computer can only produce sequences of numbers which appear to be random by some kind of irregularity. These are the pseudo-random numbers. There are several ways of producing them most of which introduced by computing modulo some fixed number m. (A similar way where the number m is replaced by a polynomial is described at the end of section 8.1).

A commonly used method which requires very little memory space is to choose a bijection f of the numbers mod m and generate a sequence

$$S : s(0), s(1), \ldots, s(n), \ldots$$

such that $s(n+1) \equiv f(s(n)) \mod m$ for all $n \geq 0$. The number $s(0)$, called the *seed* of the sequence, is not generated. One could for instance let m

be the maximal capacity of the arithmetic unit and let the machine store the last user's last number as a seed for the next user. This introduces an element of randomness.

Sequences generated by iteration has some general properties.

R. Let f be invertible mod m. Prove that all integers $n > 0$ for which $s(n) \equiv s(0)$ mod m consist of all positive multiples of the least of them, say $d > 0$, and that d divides m. The number d will be called the *order of f* mod m. (Hint. Compare the figure of section 1.1. By definition d is the smallest number $n > 0$ such that $s(n) \equiv s(0)$ mod m, or, equivalently, since f is a bijection, $s(n + j) \equiv s(j)$ mod m for all $j \geq 0$.)

LEMMA 1. *When $k > 0$ divides m, the order of f mod k divides the order of f mod m. When m is a product jk and the orders r, s of f mod j and f mod k are coprime, then rs divides the order of f mod m.*

PROOF: When d is the order of f mod k, then $s(d) \equiv s(0)$ mod k so that $s(d) \equiv s(0)$ mod m and hence d is a multiple of the order of f mod m. Hence, when $m = jk$, the order mod m of f is a multiple of the order r of f mod k and the order s of f mod j. Since r and s are coprime, rs divides the order of f mod m. This finishes the proof.

When f serves as a generator of pseudo-random numbers between 0 and m, a minimal requirement is the f be a bijection mod m, i.e. that the order of f equals m. We shall see that this is possible even for simple functions like first order polynomials $f(x) = ax + b$, provided the integers a and b satisfy certain conditions. Before passing to the statement and proof of this result, it is convenient to have the lemma of section 1.7 available. It is reproduced here as

LEMMA 2. *When $p > 2$ is a prime, when $(u, p) = 1, c > 0$ and $r > 0$ are integers, then*

(1) $$(1 + up^c)^{p^r} \equiv 1 + up^{c+r} \mod p^{c+r+1}.$$

The same holds when $p = 2$ and $c > 1$ while

(2) $$(1 + 2u)^{2^r} \equiv 1 \mod 2^{r+2}.$$

THEOREM. *Let $f(x) = ax + b$, a and b integers. The order of f mod m equals m if and only if a and b are prime to m, every prime which divides m also divides $a - 1$ and 4 divides $a - 1$ when 4 divides m.*

PROOF: A simple induction shows that

(3) $$s(n) = f^{(n)}(0) = b(a^n - 1)/(a - 1) = b(1 + a + \cdots + a^{n-1})$$

2.3 Pseudo-random numbers

for all $n > 0$. Assume that the order of f mod m is m. Then it is obvious that m and b must be coprime. If $d > 1$ divides m and a, then all terms of $s(n)$ are $\equiv 0 \mod d$ so that $s(m) \equiv 0 \mod m$ implies that d divides b which was seen to be impossible. Hence $(a, m) = 1$ so that f is invertible mod m. Hence Lemma 1 shows that the order mod q of f equals q for all primary divisors of m in its primary decomposition. Let q be such a prime and write $q = p^c$. If p does not divide $a - 1$ then (3) with $n = q$ shows that p divides $a^q - 1$ so that the order of a in $\Pr(p)$ divides q. Since the order in question also divides $p - 1$, we are left only with the case $p = 2, q = 2^c, a = 1 + 2u$ with u odd. Then (2) shows that the left side of (3) is 2^c and the right side divisible by 2^{c+1} which is a contradiction. This proves that the conditions of the theorem are necessary.

Assume now that the conditions of the theorem hold. Lemma 1 shows that in order to prove that f has order m mod m, it suffices to show that f has the order q for every primary number q in the primary decomposition of m. Hence it suffices to prove that

$$s(n) = b(a^n - 1)/(a - 1)$$

is not congruent to $0 \mod q$ when n is a power $p^s < q$ of the prime p belonging to $q = p^c$. But this follows from the formula (2) of Lemma 2, since, by hypothesis, a has the form $1 + up^c$ where $c > 0, (u, p) = 1$ and $c > 1$ when $p = 2$.

How random?

One of the functions satisfying the conditions of the preceding theorem is $g(x) = x + 1$. Its order mod m is m, but the sequence it generates, $0, 1, \ldots, m - 1$, cannot be said to be random. We shall measure the randomness of a sequence mod m,

$$S : s(0), s(1), \ldots, s(n), \ldots$$

by considering its first element $t = s(0)$ as a random variable equally distributed among the integers mod m. This makes any following term $s(n)$ a random variable $h(n, t)$. A simple way of measuring the interdependence of two such variables is to compute the mean value

$$M(j, k) = E(e(h(j, t) - h(k, t))).$$

Here $e(z) = e^{2\pi i z/m}$ maps the congruence classes mod m onto m equidistant points on the unit circle. That $M(j, k) = 0$ is then an indication that the variables $h(j, t)$ and $h(k, t)$ may be independent in some loose sense.

THEOREM. *Let $f(x) = ax + b$ satisfy the requirements of the preceding theorem. Then $M(j,k) = 0$ unless $j - k$ is a multiple of the order of a in $\Pr(m)$, in which case $h(j,t) - h(k,t)$ is constant. The order of a in $\Pr(m)$ is m/s where s is the product of the primary divisors of $a - 1$ which divide m.*

Note. The ideal situation would be that $M(j,k) = 0$ unless $j - k \equiv 0$ mod m. Our theorem shows that this is impossible. But the exceptions should be as few as possible and from this point of view, $a - 1$ should be small and a prime. The best choices are 7 and 11. In the second case m may be a large power of 10. This simple classical pseudo-random sequence, subject here to a simple statistical test, also withstands others (Knuth (1981)).

PROOF OF THE THEOREM: A simple induction shows that

$$h(n,t) = a^n t + c(n), \quad c(n) = b(1 + a + \cdots + a^{n-1}).$$

Hence

$$e(h(j,t) - h(k,t)) = e(c(j) - c(k))e(a^j - a^k)^t.$$

Since $1 + z + \cdots + z^{m-1} = 0$ or 1 according as $z \neq 0$ or $z = 1$ when $z^m = 1$, this shows that $M(j,k) = 0$ unless $a^j - a^k \equiv 0 \mod m$, i.e., since $(a,m) = 1$, $a^{j-k} \equiv 1 \mod m$. Hence $M(j,k) = 0$ unless $j - k$ is a multiple of the order of a in $\Pr(m)$, in which case $h(j,t) - h(k,t)$ is the constant $c(j) - c(k)$. By basic number theory, the order of a in $\Pr(m)$ is the product of the orders of $a \mod q$ where q runs through the primary divisors of a (in its primary decomposition) provided these orders are coprime. When m has a primary factor q with prime divisor p, then $a - 1$ has the divisor p so that $a = 1 + up^f$ with $(u,p) = 1$ and $f > 0$. By Lemma 2 above, the order of $a \mod q$ is then $\max(1, q/p^f)$ and this holds also when $p = 2$. Since all these orders are coprime and their product is m/s, the proof is finished.

Recent developments

When more complicated functions than linear ones are used to generate pseudo-random sequences, the requirements on randomness can be sharpened. One method which has been suggested is to choose a n-bit prime p, an n-bit seed s, a generator a of $\Pr(p)$ and then generate a sequence $S = (s(0), s(1), \ldots)$ where $s(0) = s$ and $s(k+1) \equiv a^{s(k)} \mod p$ when $k > 0$. Under the assumption that the inverse of $x \to a^x(p)$ cannot be computed at a bit cost which is polynomial in p, Blum and Micali (1984) have shown that, with s, p, a, chosen at random, no reasonable statistical test is able to detect a correlation between a bit of S and the following bit.

Their paper gives the precise formulation of this and other similar results and references to some of the recent literature.

Literature

The bible of arithmetic and other algoritms is Knuth (1973), (1981). The main source for this chapter has been Aho-Hopcroft-Ullman (1974).

CHAPTER 3

Abstract algebra and modules

This chapter introduces the basic notions of abstract algebra as they are used in mathematics today. The notions are illustrated in detail with a study of *modules*, also called *abelian groups*, and defined as sets equipped with addition and subtraction (additive module) or commutative multiplication and division (commutative or abelian group). From an abstract point of view these two are the same, but it is useful to distinguish between them in practice.

The modules are very useful for a first acquaintance with the terminology of algebra, in particular direct sum, quotient, and morphisms. Most examples are from number theory where the additive module of congruence classes mod m and the group of congruence classes mod m which are coprime to m are all important.

The chapter ends with the structure theorem for finitely generated modules. This requires more patience and imagination of the reader than the preceding text.

The next chapter deals with some applications of module theory, in particular the finite Fourier transform and the fast Fourier transform.

3.1 The four operations of arithmetic

The point of algebra is to have abstract models which fit into many special cases. The four operations of arithmetic transplanted into an abstract landscape are such models. Here follows the precise definitions. All of them are described in terms of an unspecified set A with elements $a, b, c, \ldots$. The text below can be used as a reference when the words addition etc. are used in the sequel.

Addition

To every pair of elements a, b of A there is a unique third element, called the sum of a and b and denoted by $a + b$, with the following properties,

$$a + b = b + a \text{ (the commutative law)}$$

and

$$(a + b) + c = a + (b + c) \text{ (the associative law)}$$

for all a and b.

3.1 The four operations of arithmetic

Note. The sign $+$ is used for simplicity. In principle, any other sign could do, for instance $\wr, \diamond, |$ and so on. The sum $a + b$ is also a function of a and b with values in A. If we denote it by $f(a,b)$, the rules above read $f(a,b) = f(b,a)$ and $f(f(a,b),c) = f(a,f(b,c))$. The same remark applies to the ensuing definitions.

Subtraction

The set A has an addition and there is one element of A called *zero* such that
$$a + 0 = 0 + a = a$$
for all a in A. To every a there is an element called the *opposite* or *additive inverse* of a and denoted by $-a$ such that
$$a + (-a) = 0$$

Note. There is just one zero element for if 0 and $0'$ have the properites of a zero, then $0 = 0 + 0' = 0'$. Similarly, there is just one opposite b of a for if there were another one, c, then $c + 0 = c + (a+b) = (c+a) + b = 0 + b = b$.
Note. The sum $a + (-b)$ is also written $a - b$.

Multiplication

To every pair of elements a, b of A there is a unique third element of A called the *product* of a and b and denoted by ab such that
$$(ab)c = a(bc) \text{ (the associative law)}$$
for all a, b, c in A.

Division

The set A has a multiplication and there is in A an element called the *unit* or one and denoted by 1 or e such that
$$a1 = 1a = a$$
for all a in A. To every non-zero a in A there is one element called the inverse of a and denoted by a^{-1} such that
$$a^{-1}a = aa^{-1} = 1$$

R. Prove that any two units are the same and also any two inverses.

Distributivity

The set A has both addition (or addition and subtraction) and multiplication connected by the left and right distributive laws
$$a(b+c) = ab + ac, \ (b+c)a = ba + ca$$
for all a, b, c in A.

Note. Multiplication is perhaps the most primitive law. Composition of functions $f(x)$ from a set X to itself defined by $(fg)(x) = f(g(x))$ for all x in X satisfies the associative law and hence is a multiplication. Composition of functions is only rarely commutative. For instance, if $f(x) = x + b$ and $g(x) = ax$ with a, b, x real, then $(fg)(x) = ax + b$ but $(gf)(x) = ax + ab$.

R. Suppose that A has addition and subtraction connected with multiplication by the distributive laws. Show that $a0 = 0a = 0$ for all a in A. Show also that $(-a)(-b) = ab$ for all a and b.

Terminology

For sets provided with one or several of the abstract arithmetic operations, the following list of terms is used.

Monoid or *semigroup*: multiplication
Group: multiplication and division
Module: addition and subtraction
Ring: module with multiplication and the distributive laws
Division ring: ring where every element $\neq 0$ has a multiplicative inverse
Field: division ring with commutative multiplication

There are examples of all these notions in number theory with the natural arithmetic operations. The rational, real and complex numbers $\neq 0$ constitute three groups under multiplication. But the integers $\neq 0$ do not form a group under multiplication since $1/a$ is not an integer when a is an integer $\neq \pm 1$. The set Zm with m a fixed integer is both a module and a ring in the abstract sense and has a commutative multiplication. It has a unit only when $m = \pm 1$. Z is not a field since ± 1 are the only invertible elements. The rational numbers Q on the other hand constitute a field since rational numbers $\neq 0$ have multiplicative inverses. The real numbers R and the complex numbers C are also fields.

R. Verify that the invertible elements of a monoid with a unit form a group. Special case: the elements $\neq 0$ of a division ring form a multiplicative group.

From the abstract point of view there is no difference between a module and a commutative group. We just have to write addition and subtraction instead of multiplication and division and vice versa. In this way, a

3.1 The four operations of arithmetic

multiplicative unit and the zero for addition correspond to each other. The distinction between the two concepts is just traditional and terminologically convenient.

Number theory and arithmetic mod m offer a non-trivial example of a commutative ring, namely the set Z_m of congruence classes

$$C(x) = x + mZ$$

mod m. Here x is any integer, said to *represent* the class $C(x)$. Addition, subtraction and multiplication in this set are defined by the formulas

(1) $$C(x) \pm C(y) = C(x \pm y),\ C(x)C(y) = C(xy).$$

In order to verify that these definitions make sense and satisfy the axioms we first note that x and y belong to the same class $C(z)$ if and only if $x - z \equiv 0\ (m)$ and $y - z \equiv 0\ (m)$ from which follows that $x - y \equiv 0\ (m)$. Now if $x \equiv x'\ (m)$ and $y \equiv y'\ (m)$, we know that $x \pm y \equiv x' \pm y'\ (m)$ and $xy \equiv x'y'\ (m)$. Hence the right sides above do not depend on the choice of representatives of the classes $C(x)$ and $C(y)$. The axioms are now easy to verify. Since $(x + y) + z = x + (y + z)$ for integers, we have $C((x + y) + z) = C(x + (y + z))$ and hence, by the rules (1),

$$C(x + y) + C(z) = C(x) + C(y + z)$$

and again by the rules (1),

$$(C(x) + C(y)) + C(z) = C(x) + (C(y) + C(z)).$$

The other axioms are verified in the same way. It is clear that $C(0)$ and $C(1)$ are the zero and unit of Z_m. We have now proved most of

THEOREM. *The set Z_m of congruence classes mod m constitutes a ring. It is a field if and only if m is a prime.*

PROOF: When $m = p$ is a prime and a is not divisible by p, we know that there is an integer b such that $ab \equiv 1\ (p)$ which means that $C(a)C(b) = C(1)$.

R. Prove that Z_m is not a field when m is not a prime.

The set of invertible elements of Z_m constitute a commutative group denoted by $Z_m{}^*$. It consists of all classes $C(x)$ with x prime to m. In fact, then and only then is there an integer y such that $xy \equiv 1\ (m)$. This group has $\varphi(m)$ elements and may or may not be cyclic, i.e., consist of all powers of a suitable element. The cases when it is cyclic are listed in section 1.7. The general structure theorem in section 3.5 below applies to this group, but does not describe all details. They depend on the number m in a complicated way.

Note. Congruence classes mod m are sometimes written in a simpler way, for instance as (x) or $\bar{x}$.

Examples

The group $Z_8{}^*$, which is known under the name of Klein's four group, has four elements $C(1), C(3), C(5), C(7)$. If, for simplicity, we denote them by e, a, b, c, then e is the unit, all squares are 1 because $3^2 \equiv 1$ (8) etc. and $ab = ba = c$ (because $3 \cdot 5 - 7 \equiv 0$ (8)). This describes the group completely, for multiplying $ab = c$ by a and b we get $b = ac$ and $a = bc$. This describes all 16 products. The group has the following multiplication table:

*	e	a	b	c
e	e	a	b	c
a	a	e	c	b
b	b	c	e	a
c	c	b	a	e

where an element of the inner 3×3 matrix is the product of the two outer elements in the same row and column.

R. Compute the corresponding table when $m = 12$.

The purpose of the next section is to present some standard constructions of abstract algebra when applied to the simplest case of modules and commutative groups.

3.2 Modules

An abstract *module* is a non-empty set $M = \{a, b, c, \ldots\}$ with addition and subtraction in the abstact sense. Let 0 be the zero of a module M. (There is no risk of confusion with the ordinary zero.) If we put

$$0a = 0, \quad ma = a + \cdots + a, \; (m \text{ terms}), \quad -ma = m(-a),$$

where $m > 0$ is an integer and a any element of M, it is easy to check that all na with n in Z and a in M are elements of M for which we have the following rules

$$(m+n)a = ma + na, \quad m(a+b) = ma + mb, \quad m(na) = (mn)a,$$

for all m and n in Z and a and b in M. We express this by saying that Z *operates* on M, or that M is a Z-*module*.

3.2 Modules

For a group, the rules above, namely

$$a^m a^n = a^{m+n}, \ (ab)^m = a^m b^m, \ (a^n)^m = a^{nm},$$

are the familiar rules for exponentiation.

Submodules

A non-empty subset N of a module M is called a *submodule* if $a - b$ is in N when a and b are.

R. Prove that every submodule is itself a module. (Hint: If N contains an element a, it also contains $a - a = 0$, then $0 - a = -a$, etc.)

Any finite subset $\{a, b, \ldots, c\}$ of a module M generates a submodule, namely the set

$$Za + Zb + \cdots + Zc$$

of sums $ma + nb + \cdots + rc$ of integral multiples of $a, b, \ldots, c$. The elements $a, b, \ldots, c$ are called the *generators* of this submodule.

Cyclic modules

A module Za with just one generator a is said to be *cyclic*. If its elements

$$\ldots - 2a, -a, 0, a, 2a, \ldots$$

are all different, it behaves just like the integers under addition and subtraction. Under all circumstances, the integers r such that $ra = 0$ form a module of integers, for if $ra = 0$ and $sa = 0$, then $(r - s)a = 0$. We know that such a module has the form Zm, $m \geq 0$. If $m = 0$, all the multiples of a are different. Otherwise

$$0, a, \ldots, (m-1)a$$

are all the elements of Za and they are all different. Addition and subtraction in Za are performed under the condition that $ma = 0$. This determines these operations completely; we have $ra = sa$ if and only if $r \equiv s \ (m)$. This module is said to be *cyclic* of order m, where 'order' means the number of elements. When $a = 0$, Za contains only the element 0. The elements of a cyclic module of order m behaves exactly as the congruence classes mod m under addition and subtraction.

The basis of additive number theory as it is presented in Chapter 1 is that every module of integers is cyclic. Using this fact we can prove an abstract version.

THEOREM. *Every submodule N of a cyclic module M is cyclic. When the order of M is finite, the order of N divides that of M.*

PROOF: N consists of multiples ka of the generator of M. Since N is a submodule, $(k - j)a = ka - kj$ is in N when ka and ja are. It follows that all k such that ka is in N constitute a module T of integers and hence $T = Zn$ for some integer $n \geq 0$. When $n = 0$, $T = 0$, when $n = 1$, $T = M$. In general na generates N so that N is cyclic. When M has finite order m, then $ma = 0$ so that m belongs to the module T. It follows that n divides m, $m = nr$ and that $0, na, \ldots, (r - 1)na$ are the elements of N and that they are all different. Hence r is the order of N. The proof is finished.

The properties of coprime integers produce the next result.

THEOREM. *Let A and B be cyclic submodules of a module M and suppose that the orders m and n of A and B are coprime. Then A+B is a cyclic module of order mn.*

PROOF: The elements of $A + B$ have the form $a + b$ with a in A and b in B. Hence $A + B$ is a module. Let a be the generator of A and b that of B. Then $c = a + b$ has the order mn. In fact, if $rc = 0$, then $0 = ra + rb$ so that $0 = mra = -mrb$. It follows that n divides r. Similarly, m divides r. Hence r is a multiple of mn which is also the order of c. Since $A + B$ has at most mn elements, c generates $A + B$. This finishes the proof.

Group notation

For commutative groups, the notion of submodule corresponds to subgroup. A part H of a group G is called a subgroup if not empty and ab^{-1} is in H when a and b are. It follows that H is a group. A subset $\{a, b, \ldots\}$ of G is said to generate G if every element of G is a product of powers of these elements. A cyclic group G is one generated by a single element a and then it consists of all powers of a. When G has a finite number n of elements, these elements are

$$e = a^0, a, a^2, \ldots, a^{n-1}.$$

As for cyclic modules one proves that the subgroups of a cyclic group G are cyclic and that the order of a subgroup divides the order of G.

We have seen in section 1.4 that the group of congruence classes mod m which are coprime to m is cyclic only when m is a power of a prime > 2, when m is twice such a number and when m is 2 or 4, but in no other case. This class of groups will serve to illustrate the structure theorem for finite groups proved later in this chapter.

3.2 Modules

Exercise

How many elements of order 5 are there in a cyclic module of order 20? How many in one of order 10? Show that Klein's four group has three proper subgroups, i.e., subgroups not equal to the unit element or the entire group.

Quotients

The constructions which follow imitate the construction of Z_m considered as a module.

Associated with any submodule N of a module M is the *quotient* module M/N whose elements are the *cosets*

$$(a) = a + N$$

of N. (We could also use a function notation $f(a) = a + N$ for the cosets.) The cosets inherit a module structure from M if we let $(0) = N$ be the zero of M/N, $(-a)$ the opposite of (a) and define addition by

$$(a) + (b) = (a+b).$$

This statement is easy to prove and we shall do it.

A *partition* of a set X is a collection of subsets Y of X, no two of which have an element in common, with the property that they cover X: every x in X belongs to some Y in the collection. The reader should visualize partitions of finite sets and note that there is no condition on the nature of the subsets or their number.

We first show that the cosets (a) form a partition of M. In fact, any a is in its coset (a) and if two cosets (a) and (b) have an element in common, $a+c = b+d$, then $a+N = b+d-c+N = b+N$ since, obviously, $e = d-c$ is in N and $e + N = N$ for any e in N.

Since $(0) + (a) = (0 + a) = (a)$ for all a, (0) is the zero of M/N, and since $(-a) + (a) = (0)$, $(-a)$ is the opposite of (a). Next we have to show that $(a+b)$ is a function of (a) and (b). In fact, $(a) = (a')$ and $(b) = (b')$ if and only if $a - a'$ and $b - b'$ are in N and then

$$(a'-b') = a'-b'+N = (a-b)+(a'-a)+(b'-b)+N = a-b+N = (a-b).$$

The proof that the addition in M/N is associative is left to the reader.

Examples

When Za is infinite cyclic and $m > 0$ is an integer, the quotient Za/mZa is a cyclic module with m elements generated by the coset $(a) = a + mZa$. When $a = 1$ is in Z we get the important module

$$Z_m = Z/mZ$$

of the familiar congruence or residue classes mod m. The quotient module $M/0M$ is of course the same as M itself for any module M.

In group notation, quotients amount to the following. If G is a commutative group and H is a subgroup, the elements of the quotient G/H are the cosets aH of elements a of G. Carrying over the proof above, one sees that the cosets partition G and that

$$(aH)(bH) = abH$$

turns the set of cosets into a group.

Exercise

Let $G = \{e, a, b, c\}$ be Klein's four group and H the subgroup generated by a. List the cosets of H and establish a multiplication table for the group G/H.

Direct sums of modules

When M and N are modules, we can form their *direct sum* $M \oplus N$ consisting of pairs (a, b) with a in M and b in N subject to the following rules. The zero of $M \oplus N$ is the pair of zeros of M and N, the opposite of (a, b) is $(-a, -b)$, addition is performed according to the formula

$$(a, b) + (c, d) = (a + c, b + d)$$

(componentwise addition) and similarly for subtraction.

The same construction can be expressed somewhat differently and then also generalized. Let X be a set and suppose that there is a module M_x for every $x \in X$. Then all functions f from X such that $f(x)$ is an element of M_x form a module where the zero is the zero function, the opposite of $f(x)$ is $-f(x)$ and addition is performed by the rule for adding functions,

$$(f + g)(x) = f(x) + g(x).$$

In particular, X can be the set $\{1, 2, \ldots, n\}$ and we get the direct sum of n modules.

R. Let M and N be modules with m and n elements respectively. Prove that the direct sum $M \oplus N$ has mn elements.

Note that a direct sum is not the same as an ordinary sum. Suppose for instance that $a = b$. Then $Za + Zb = Za$ is generated by one element in contrast to the direct sum $Za \oplus Za$ which has two generators, $(a, 0)$ and $(0, a)$.

$y \in H$. Multiplication is done componentwise,
$$(x,y)(u,v) = (xu, yv).$$
It follows that multiplication is commutative (since it is so in G and H). The unit of the product is (e, f) where e is the unit of G and f is that of H. The inverse of (x, y) is (x^{-1}, y^{-1}).

Exercise

How many elements of order 5 are there in the direct product of two cyclic groups of order 20 and 10? (Answer: 7)

3.3 Module morphisms. Kernels and images

Together with modules it is important to consider also maps between them. A map $f : M \to M'$ from one module $M = \{a, b, c, \dots\}$ to another one $M' = \{a', b', c', \dots\}$ is said to be a *homomorphism* or, more precisely, a module *morphism* if it preserves the module structure in the sense that
$$f(a + b) = f(a) + f(b)$$
for all a and b in M. An *endomorphism* is a homomorphism from M to itself. A bijective homomorphism is called an *isomorphism* and, if $M = M'$, an *automorphism*.

R. Prove that any two cyclic modules A and B of the same order are isomorphic. (Hint. If a and b are the generators, verify that $na \to nb, n$ any integer, is an isomorphism.) Construct a homomorphism $A \to B$ when the order of B divides that of A. (Hint. If the order of A is k times that of B, consider the map $na \to nkb$.)

Examples

If M is a direct sum $L \oplus N$ with elements (a, b), both maps $(a, b) \to a$ and $(a, b) \to b$, called *projections* to L and N respectively, are homomorphisms. Also, if N is a submodule of a module M and M/N is the quotient with elements $(a) = a + N$, the map $f(a) = (a)$ is, by its very definition, a homomorphism from M to M/N. The map $n \to na$ from Z to any cyclic module Za is a homomorphism; if Za is infinite, it is an isomorphism.

The following simple theorem shows that a simple sum is sometimes isomorphic to the direct sum.

THEOREM. *The sum $A + B$ of two submodules A and B of a module M is isomorphic to the direct sum $A \oplus B$ if $A \cap B = 0$.*

PROOF: If a, a' are in A and b, b' in B and $a + b = a' + b'$ then $a - a' = b' - b$ belongs to $A \cap B$ and hence vanishes. It follows that the pair (a, b) is a function of the sum $a+b$. The map $a+b \to (a,b)$ from $A+B$ to the direct sum $A \oplus B$ is a module morphism, for

$$(a+b) - (a'+b') = a - a' + b - b' \to (a-a', b-b') = (a,b) - (a',b').$$

R. If A and B are cyclic submodules of coprime orders m and n, prove that $A \cap B = 0$. Prove also that if a and b are the generators of A and B, then $a+b$ has order mn. Prove that

$$(ra, sb) \to (rm' + sn')(a+b)$$

where $mm' + nn' = 1$ is an isomorphism from $A \oplus B$ to the cyclic module generated by $a+b$.

Exercise

Let M and N be modules. Show that the direct sums $M \oplus N$ and $N \oplus M$ are isomorphic.

Let $f : M \to M'$ be a module morphism. Then the *image*

$$\mathrm{im} f = f(M)$$

of f is a submodule of M'. To see this, note that if $f(a)$ and $f(b)$ are images of elements in M, so is $f(a) - f(b) = f(a-b)$.

The set of elements in M which are mapped by f to the zero $0'$ of M' is called the *kernel* of f, ker f. It is a submodule of M, for if $f(a) = f(b) = 0$, then $f(a-b) = 0'$. We provide a picture of the situation below.

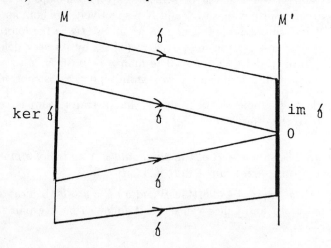

3.4 The structure of finite modules

R. Prove that a module morphism is injective if and only if its kernel vanishes.

R. Prove that the kernel of the homomorphism $a \to (a)$ from M to M/N is the module N.

R. Let f be a homomorphism from the direct sum $Za \oplus Zb$ of two cyclic modules. Show that $f(a,0)$ and $f(0,b)$ generate im f. Show that these elements of the receiving module can be given arbitrarily when Za and Zb are infinite cyclic modules but not otherwise.

The following theorem is an important terminological exercise. It will appear again and again in different disguises. Stating it here in a simple situation, we hope to be able to treat it more lightly in connection with rings and groups.

MODULE MORPHISM THEOREM. *When $f : M \to N$ is a module morphism, the image of f is isomorphic to the quotient $M/\ker f$.*

PROOF: The elements of the quotient $M/\ker f$ are the cosets

$$(a) = a + \ker f.$$

The coset (a) is mapped by f into $f(a)$, since $f(a+b) = f(a) + f(b) = f(a)$ when $f(b)$ vanishes in N. It is also clear that $f((a) - (b)) = f((a)) - f((b))$. In fact, this says nothing more than $f(a - b) = f(a) - f(b)$. The map f is surjective to im f and if $f((a))$ is zero, then $(a) = \ker f$, which is precisely the zero of $M/\ker f$.

R. Let Za be a cyclic module with m elements. Show that Za and Z/mZ are isomorphic.

R. Let $L \subset N \subset M$ be three modules. Note that N/L can be considered as a submodule of M/L. Show that $(M/L)/(N/L)$ is isomorphic to M/N. (The second homomorphism theorem. Hint. Use the module morphism theorem.)

The modules Hom(M, N)

Let M and N be two modules and $f, g : M \to N$ two module morphisms. Define their sum $f + g$ by

$$(f + g)(x) = f(x) + g(x).$$

This gives the structure of a module to the set $\text{Hom}(M, N)$ of all homomorphisms from M to N.

R. Verify this statement.

R. Let M and N be cyclic modules with m and n elements respectively. Show that $\mathrm{Hom}(M,N) = 0$ when m and n are coprime. (Hint. m annihilates the image of M in N.)

Group morphisms

To carry over this section to groups is just a matter of notation. A homomorphism from a group G to another one H is a map $f : G \to H$ such that $f(xy) = f(x)f(y)$. The kernel of f consists of the elements of G for which $f(x)$ is the unit of H. The image of f is im $f = f(G)$.

R. Prove that the kernel and the image of a group homomorphism f from G to H are subgroups of G and H respectively. Prove that the image $f(G)$ of G is isomorphic to the quotient $G/\mathrm{ker} f$.

R. Prove that Klein's four group $\{e, a, b, c\}$ is isomorphic to the direct product of the subgroups generated by a and b (or any two of the elements a, b, c).

Exercises

1. Let M and N be cyclic modules with 8 and 12 elements respectively. How many elements does $\mathrm{Hom}(M,N)$ have?
2. Let M and N be finite modules. Show that the maximal order of $\mathrm{Hom}(M,N)$ divides the maximal order of N.
3. Let M and N be cyclic modules with 18 and 8 elements respectively. How many elements of $\mathrm{Hom}(M,N)$ are surjective and how many are injective?
4. Let M be a cyclic module with 4 elements and put $N = M \oplus M$. Show that $\mathrm{Hom}(N,N)$ has 256 elements of which 96 are isomorphisms.

3.4 The structure of finite modules

This section, which is more than a terminological exercise, will show how finite modules are built from cyclic ones. Changed to multiplicative notation, the results also apply to commutative groups.

We know that the order of any of its elements divides the order of a finite cyclic module. For use later we shall now prove a stronger statement. First

LEMMA. *Let a and b be elements of orders m and n in a module M. If m and n are coprime, the order of $a + b$ is mn. If n does not divide m, then the module $Za + Zb$ has elements of order $> m$.*

Note. The proof is the same as for the last lemma of section 1.3 but we give it here again or the convenience of the reader.

PROOF: Let r be the order of $a+b$. Then $ra+rb = 0$ so that $rna = rmb = 0$. It follows that m divides rn and that n divides rm. Hence both n and m

divide r. This proves the first part of the lemma. To prove the second part, let us write m and n as products of powers of primes. Since n does not divide m, there is a prime p and a power q of p such that q but not pq divides m but pq divides n. Write $m = m'q, n = n'pq$. Then qa has order m' and $n'b$ has order pq. Since m' and pq are coprime, $qa + n'b$ has order $m'pq = mp > m$. This finishes the proof.

The *maximal order* of a module M is defined to be the maximum of all the orders of the elements of M. The following important result follows immediately from the lemma.

LEMMA 1. *The order of any element of a finite module M divides its maximal order.*

PROOF: Let m be the maximal order and a an element of M with the order m. Take any element $b \in M$ and apply the lemma.

Before proving the structure theorem for finite modules, we shall list and prove two more lemmas.

LEMMA 2. *Let C be a cyclic module of order m and suppose that d divides m, $m = dd'$. Then the equation $dx = y$ has a solution if and only if $d'y = 0$.*

PROOF: If $dx = y$, then $0 = dd'x = d'y$. Conversely, let z be a generator of C. Since $d'y = 0$, y is a multiple of dz, $y = sdz$, and it suffices to take $x = sz$.

LEMMA 3. *Let B be a submodule of a finite module A and let $f : A \to B$ be a morphism from B to a cyclic module of order m. Suppose that $mx = 0$ for all x in A. Then f can be extended to a morphism $F : A \to B$.*

PROOF: It is obviously enough to prove that f extends to some submodule of A of which B is a proper part. Let y be an element of A outside B. Then $B' = A + Zy$ is such a submodule. All integers k for which ky is in B form a module in Z and hence they are all multiples of an integer $d > 1$ and, since $my = 0$ by assumption, we have $m = dd'$ for some integer d'.

If a morphism F existed with the desired properties and $F(y) = c$, we should have
$$F(x + ny) = f(x) + nc$$
for all integers n. We shall see that there is a c such that this formula defines the desired morphism from B' to C. Since $f(dy)$ is in C and
$$d'f(dy) = f(dd'y) = f(my) = 0,$$
Lemma 2 shows that $f(dy) = dc$ for some c in C. This c will be our choice. Then, if $x + ny = x' + n'y$, $x - x' = (n' - n)y$ is a multiple kdy of y and we

get

$$f(x) + nc - f(x') - n'c = f(x - x') + (n - n')c = f(kdy) - kdc = 0.$$

R. Verify that F is a module morphism.

We are now ready for

THE STRUCTURE THEOREM FOR FINITE MODULES. *Every finite module A of order > 1 is the direct sum of non-trivial cyclic submodules $A_1, \ldots, A_n$ which can be chosen so that the order of each module divides the order of the preceding one.*

Note. It is shown in exercises below that these orders (but not the modules themselves) are then uniquely determined by A.

Note. Groups. The theorem carries over to commutative groups and direct products. Examples are provided by the group $\Pr(m)$ of congruence classes mod m which are coprime to m. We know from section 1.7 which of these groups are cyclic. The simplest non-cyclic group, Klein's four group, occurs when $m = 8$. If is a group of order 4 and the direct product of two subgroups of order 2. The group $\Pr(100)$ has $\phi(100) = 40$ elements. One finds that 3 generates a subgroup of order 20 and that the group is a direct product of the groups generated by the congruence classes of 3 and -1. There are also more complicated examples but no easy way to find the decompositions. The theorem is a pure existence result.

Note. It will be shown in an appendix to this chapter that every finitely generated module A is a direct sum of a finite module T and a direct sum F of infinite cyclic modules. The module T, which consists of all elements x of A for which $nx = 0$ for some integer n, is called the torsion submodule of A and F the free part.

PROOF: When A itself is cyclic, we are done. If A is not cyclic, its maximal order m is a proper divisor of the order $|A|$ of A. Choose an element x of A of maximal order m and put $B = \mathbb{Z}x$. We shall see that there is a submodule C of A such that $A = B \oplus C$. If we use induction, this proves the theorem, for $|C| < |A|$ and, by Lemma 1, the order of any element of C divides m.

To show that C exists, we extend the identical map from B to itself to a morphism $F : A \to B$. Lemma 3 shows that this is possible, since B is cyclic and $mA = 0$. We now claim that $\ker F$ will do as C. To show this, take a z in A and write

$$z = F(z) + (z - F(z)).$$

Here the first term on the right is in B and the second is in ker F, for F is the identity on B and hence

$$F(z - F(z)) = F(z) - F(F(z)) = F(z) - F(z) = 0.$$

R. Complete the proof by verifying that B and ker F have only 0 in common.

Exponents and types

The maximal order of a module A can also be described as the least integer m that annihilates A, $mA = 0$. The number m is called the *exponent* of A and the succession of exponents $m_1, \ldots, m_n$ of the modules $A_1, \ldots, A_n$ the *type* of A. In the form of exercises we shall show below that two finite modules are isomorphic if and only if they have the same type.

R. Show that a finite module is cyclic if and only if its order and type are the same.

The divisibility condition for types implies restrictions

R. Show that (50) and $(10, 5)$ are the only types of a module of order 50 and that (16), $(8, 2)$, $(4, 4)$, $(4, 2, 2)$, $(2, 2, 2, 2)$ are the only possible types when the order is 16.

R. Show that if a number d divides the order of a finite module, then it has a submodule of order d. (Hint. Use the theorem.)

R. Let x and y be two elements of order m and n of a module. Show that $x + y$ has the order mn only if m and and n are coprime.

R. Show that the direct sum of two cyclic modules is cyclic if and only if the orders are coprime.

Uniqueness

The following exercises give a proof that two isomorphic finite modules have the same type. Since cyclic modules of the same order are isomorphic, the converse is trivial.

When A is a module and $m > 0$ an integer, let $A(m)$ be the set of elements of A annihilated by m, $mx = 0$.

R. Let p be a prime. Show that

$$A(p) \subseteq A(p^2) \subseteq A(p^3) \subseteq \ldots$$

is an increasing sequence of submodules of A and that the order of any element of any of them is a power of p.

Modules which are annihilated by a power of a prime p are called p-modules. Since we are dealing with finite modules, the chain of p-modules above becomes constant for a certain power of p, p^n. It is clear that n is the same for isomorphic modules. This remark is the basis for all that follows.

R. Show that $A(mn)$ is the direct sum of $A(n)$ and $A(m)$ when m and n are coprime.

R. Let A be a finite module of order m. Let $q_1,\ldots,q_n$ be the primary factors of m and let $p_1,\ldots,p_n$ be the corresponding primes. Show that A is the direct sum of the modules $B_k = A(q_k)$ and that B_k is the maximal p_k-submodule of A.

R. Let B be a module whose type is the following powers of a prime p,

$$(n_1,\ldots,n_1,\ldots,n_t,\ldots,n_t)$$

where $n_1 > n_2 > \ldots$ and n_i is repeated r_i times. Show that $|B(p^j)|$ is p to the power $f(j)$, where

$$f(j) = r_1 \min(n_1, j) + \cdots + r_t \min(n_t, j)$$

is an increasing piecewise linear function of j. Show that

$$0 = n_{t+1} < n_t < \cdots < n_1$$

are the successive break points in the graph of f and that the graph has slope $r_1 + \cdots + r_i$ between n_{i+1} and n_i. It follows that this function defines the type of B. Hence isomorphic p-modules have the same type.

R. Show that if a finite module has the type

$$(m_1,\ldots,m_t),$$

and $q_i = p^{n_i}$ is the largest power of p dividing m_i, then the type of $A(q_i)$ is $(n_1,\ldots,n_t)$. Hence the type of A is determined by the type of its maximal p-module and hence is the same for isomorphic modules.

Exercises

1. Prove that $\Pr(75)$ has the type $(20, 2)$ and write the group explicitly as a direct product of two cyclic groups. (Hint. $|\Pr(75)| = 40$ and we know that $\Pr(75)$ is not cyclic. Try 2 for an element of maximal order.)

2. What is the type of $\Pr(100)$?

3.5 Appendix. Finitely generated modules

Let A be a module. We say that A is *finitely generated* if there is a finite set of elements $a_1,\ldots,a_k$ of A such that every a in A can be written

$$a = n_1 a_1 + \cdots + n_k a_k$$

for some integers $n_1,\ldots,n_k$. The elements $a_1,\ldots,a_k$ are called *generators* of A. If A is finitely generated, we say that it is *free* (on $a_1,\ldots,a_k$) if

$$n_1 a_1 + \cdots + n_k a_k = 0 \Rightarrow n_1 = \cdots = n_k = 0.$$

When A is a module, we denote by $T(A)$ the submodule of A consisting of all elements a of A such that $na = 0$ for some integer $n \neq 0$. Such elements are called *torsion elements* and $T(A)$ is called the *torsion submodule* of A.

R. Verify that $T(A)$ is a module.

We are going to prove that every finitely generated A module can be written $A = F \oplus T(A)$, where F is free and finitely generated. We begin with a

LEMMA. *A submodule of a finitely generated free module is free.*

PROOF: Let A be a finitely generated free module and B a submodule. We will use induction over the number of generators of A. We leave it to the reader to verify that a submodule of a free module generated by one element is free. Let $a_1,\ldots,a_k$ be a set of generators of A such that

$$A = \mathbb{Z}a_1 \oplus \cdots \oplus \mathbb{Z}a_k.$$

Define a morphism $p: A \to \mathbb{Z}a_k$ by

$$p(n_1 a_1 + \cdots + n_k a_k) = n_k a_k.$$

The kernel B' of the restriction of p to B is contained in

$$\mathbb{Z}a_1 \oplus \cdots \oplus \mathbb{Z}a_{k-1},$$

hence is free by the induction hypothesis. If $p(B) = 0$, then we are finished. Otherwise $p(B)$ is generated by some na_k, $n \neq 0$ (since $p(B)$ is contained in the cyclic module $\mathbb{Z}a_k$). Suppose that $p(b) = na_k$. We claim that $B = B' \oplus \mathbb{Z}b$. For if x is in B and $p(x) = n'na_k$, then $p(x - n'b) = 0$, and so $x - n'b$ is in B'. Furthermore, if x is in $B' \cap \mathbb{Z}b$, say $x = n'b$, then

$0 = p(x) = n'na_k$, and $n' = 0$. Hence $x = 0$ and $B' \cap Zb = 0$. This finishes the proof.

It follows from the proof that there is a finite set of generators $b_1, \ldots, b_l$ such that B is free on $b_1, \ldots, b_l$ and $l \leq k$. It is not difficult to prove that the number l is uniquely determined by B. For let $c_1, c_2, \ldots$ be another set (not necessarily finite) of generators on which B is free and assume that it has at least m elements. It is enough to show that $m \leq l$. If p is a prime, then B/pB is isomorphic to the direct sum of l copies of Z/Zp, whence $|B/pB| = p^l$. But B/pB also contains a submodule isomorphic to m copies of Z/Zp, so $p^m \leq p^l$. The number l is called the *rank* of B.

THEOREM. *If A is a finitely generated module, then $A = F \oplus T(A)$, where F is finitely generated and free.*

PROOF: Let $A' = A/T(A)$ and denote the quotient map $A \to A'$ by $a \to \bar{a}$. If $A' = 0$, then we are finished. Otherwise take a maximal set of elements $a_1, \ldots, a_k$ of A such that
$$n_1 \bar{a}_1 + \cdots + n_k \bar{a}_k = 0 \Rightarrow n_1 = \cdots = n_k = 0.$$
Let $\bar{A} = Z\bar{a}_1 + \cdots + Z\bar{a}_k \subseteq A'$. Then $\bar{A}$ is free. If $\bar{a}$ is in A', then there are integers $n \neq 0, n_1, \ldots, n_k$ such that
$$n\bar{a} + n_1 \bar{a}_1 + \cdots + n_k \bar{a}_k = 0.$$
Since A' is finitely generated (why?), there is an integer $m \neq 0$ such that $m\bar{a} \in \bar{A}$ for all $\bar{a} \in A'$, or $mA' \subseteq \bar{A}$. But A' is torsion-free, so the map $\bar{a} \to m\bar{a}$ is injective, and A' is isomorphic to a submodule of the free module $\bar{A}$. Hence A' is free by the lemma.

Define
$$F = Za_1 + \cdots + Za_k.$$
Then F is free. In fact, if $n_1 a_1 + \cdots + n_k a_k = 0$, then $n_1 \bar{a}_1 + \cdots + n_k \bar{a}_k = 0$ and $n_1 = \cdots = n_k = 0$. We claim that A is the direct sum of F and $T(A)$. For if a is in A, then $\bar{a} = n_1 \bar{a}_1 + \cdots + n_k \bar{a}_k$ for some integers $n_1, \ldots, n_k$. Hence $a - n_1 a_1 - \cdots - n_k a_k \in T(A)$ since its image in A' vanishes. It remains to prove that $F \cap T(A) = 0$. Suppose that $a \in F \cap T(A)$. Write $a = n_1 a_1 + \cdots + n_k a_k$. Then $0 = \bar{a} = n_1 \bar{a}_1 + \cdots + n_k \bar{a}_k$ and $n_1 = \cdots = n_k = 0$ since A' is free. Hence $a = 0$. The proof is finished.

R. Prove that $T(A)$ is finite if A is finitely generated.

Literature

The study of modules first appeared in number theory. The alternative name of abelian group stems from Abel's work on algebraic equations. A reader who wants to proceed to the general theory of modules over a ring can consult any comprehensive algebra text.

CHAPTER 4

The finite Fourier transform

The Fourier transform is one of the main tools of analysis with a large number of important applications in physics, technology and statistics. In numerical applications it has to appear in discrete form as the finite Fourier transform. This transform is associated with the theory of characters of finite modules (abelian groups) which forms the first part of this chapter. The remaining parts are devoted to applications, first a proof of the quadratic reciprocity theorem and then numerical applications, in particular the method of computing the finite Fourier transform which is called the fast Fourier transform, FFT.

4.1 Characters of modules

When z is a complex number, let $|z|$ denote its absolute value. Since $|zw| = |z||w|$, all complex numbers of absolute value 1, i.e. those on the unit circle, constitute a commutative group under multiplication. When $n > 0$ is an integer, all z such that $z^n = 1$, the nth roots of unity, constitute a subgroup of order n. Generators of this group are the *primitive nth* roots of unity with the property that $z^k \neq 1$ when $0 < k < n$. An obvious nth root of unity is a number with absolute value 1 and argument $2\pi/n$. By the theory of cyclic modules, every other primitive nth root of unity is a power z^k where k and n are coprime.

Characters

A map f from a module $M = \{a, b, c, \ldots\}$ to the non-zero complex numbers with the property that

$$f(a+b) = f(a)f(b)$$

is called a *character* of M. Since $f(0) = f(0+0) = f(0)^2$, we must have $f(0) = 1$ and since $f(0) = f(a-a) = f(a)f(-a)$, $f(-a) = 1/f(a)$. Hence the image of f is a commutative group (see also the generalities of Chapter 3).

R. Show that a character is uniquely determined by its values on a set of generators. Show that a product of characters is a character.

R. The exponential function $f(x) = a^x$ with a complex $a \neq 0$ (and with a definite argument) is a character of the real numbers R. Show that it is a bounded function of x if and only if $|a| = 1$.

R. Let $k \to f(k) = a^k$ be a bounded character of the integers. Show that it is an injection if and only if a is not a rational root of unity. Show that if a is an nth root of unity, then the same map defines a character of the module $Z_n = Z/nZ$ and that the character is a bijection if and only if a is a primitive root.

R. Show that a cyclic module of order n has n different characters considered as functions from the module.

R. Let M and N be modules with characters f and g. Show that $h(a,b) = f(a)f(b)$ for all a in M and b in N defines a character of the direct sum $M \oplus N$. Show that every character has this form when M and N are cyclic. (Hint. Consider the number of different characters which have this form.)

Note. Since every module is the direct sum of cyclic ones, the last exercise shows that every module has as many characters as elements. The characters form themselves a commutative group under multiplication, $(fg)(a) = f(a)g(a), f^{-1}(a) = f(-a)$.

Note. The situation described above becomes more symmetric if we consider the characters $f, g, \ldots$ of a finite commutative group $G = \{a, b, c, \ldots\}$ defined by the property that $f(ab) = f(a)f(b)$ for all a and b. Then G has as many characters as elements and the characters form a commutative group under multiplication called the *dual* of G and sometimes denoted by $G^\star$. The group $G^\star$ also has characters, among them maps $f \to f(a)$ with a fixed in G. In fact, the equation $(fg)(a) = f(a)g(a)$ says that they are characters. Since there are as many characters as elements of G and since $f(a)f(b) = f(ab)$ where f runs through $G^\star$, we conclude that the bidual $G^{\star\star}$ of G is isomorphic to G itself.

4.2 The finite Fourier transform

LEMMA. *Let w be an nth root of unity. Then the sum*

$$\sum_{0}^{n-1} w^k$$

vanishes unless $w=1$ and then its value is n.

R. Prove this.

In the simplest case, the discrete Fourier transform maps a complex function f defined on the integers mod n to another such function Tf defined

4.2 The finite Fourier transform

by

$$(Tf)(k) = \sum_{0}^{n-1} w^{kj} f(j)$$

where w is a primitive nth root of unity. The basic property of this map is the following

THEOREM. *Putting $(T'f)(k) = Tf(-k)$, the maps TT' and $T'T$ are both n times the identity. In other words,*

$$T^{-1} = T'/n.$$

PROOF: We have

$$T'Tf(k) = \sum w^{-kj} Tf(j) = \sum \sum w^{j(p-k)} f(p),$$

where all sums run from 0 to $n - 1$. By the lemma above, the sum over j yields 0 unless $w^{p-k} = 1$. Since $n - 1 \geq p - k \geq -(n - 1)$, this can only happen when $p = k$. Hence the sum on the right equals $nTf(k)$. This proves the first part of the theorem. The proof of the second part is entirely similar.

If we put $Tf = f'$, the Fourier transform and its inverse are given by the following formulas

(1) $$f'(j) = \sum w^{jk} f(k), \quad f(k) = \sum w^{-jk} f'(k)/n$$

with summations over k and j respectively.

Notice that the right sides do not change if k and j are replaced by, respectively, $k + mn$ and $j + mn$. This is consistent with the functions f and f' being functions from Z_n to the complex numbers.

Note. We can also consider the Fourier transform T as a linear map from $\mathbb{C}^n$ to itself with the matrix $(e(jk))$ where $e(z) = \exp 2\pi i z/n$. The theorem shows that $T^2 f(k) = nf(-k)$ for all k and f and hence that $T^4 = n^2 I$ where I is the unit matrix. It follows for instance that the sum of the eigenvalues of T is its trace

$$\sum e^{2\pi i j^2/n}.$$

Such sums were considered by Gauss. We shall return to this point in section 4.3 where the finite Fourier transform is used to prove the quadratic reciprocity law.

One of the basic properties of the Fourier transform is that it converts a *convolution*

$$(2) \qquad f*g(k) = \sum_{0}^{n-1} f(k-j)g(j)$$

into multiplication of the transforms. In fact, taking the Fourier transform of the $f*g$, we have to compute the sum

$$(f*g)'(p) = \sum\sum f(j-k)g(k)w^{-jp} = \sum\sum f(j-k)g(k)w^{-p(j-k)}w^{-pk}$$

where all sums run over Z_n. Summing independently over $j-k$ and k shows that

$$(3) \qquad (f*g)' = f'g',$$

i.e. the Fourier transform of a convolution is the product of the Fourier transform of its factors.

R. Prove that

$$f*g = T^{-1}(TfTg), \quad fg = T(T^{-1}f * T^{-1}g).$$

Limit passage to Fourier series

Letting n tend to infinity, the finite Fourier transform gives the Fourier series of functions defined in the unit interval. In fact, put $w = e^{-2\pi i/n}$, write $f(k/n)$ for $f(k)$, put $n = 2m+1$ and let the sums above run from $-m$ to m. Then

$$f'(j)/n = \sum_{k=0}^{n-1} f(k/n)e^{2\pi ijk/n}.$$

Here the right side is a Riemann sum of the integral

$$a_j = \int_0^1 f(x)e^{-2\pi jx}dx$$

where j runs over all integers. A passage to the limit in the other formula (1) then gives

$$f(x) = \sum e^{2\pi ikx}a_k.$$

All our limits are of course formal but easy to justify under suitable precautions, for instance that $f(x)$ is once continuously differentiable.

4.2 The finite Fourier transform

The finite Fourier transform for direct sums

To every direct sum $M = M_1 \oplus \cdots \oplus M_r$ of finite cyclic modules, there is a finite Fourier transform

(5) $$f'(j) = \sum w^{jk} f(k)$$

with the inversion formula

(6) $$f(k) = m^{-1} \sum w^{jk} f'(j).$$

Here $j = (j_1, \ldots, j_r)$ and $k = (k_1, \ldots, k_r)$ are in M, the sums run over M and m is the order of M, $m = m_1 \ldots m_r$ with m_s the order of M_s. The characters w^{jk} are products of characters

$$w_s^{pq}$$

with $p = j_s, q = k_s$ and w_s a primitive n_sth root of unity. As for cyclic modules, the proof relies on the fact that $\sum_k w^{jk} = 0$ unless all j_s vanish in which case the sum is m.

R. Prove this statement. Prove also that the Fourier transform of a convolution is the product of the Fourier transforms of its factors.

Note. The general form of the Fourier transform for a finite module $M = \{a, b, c, \ldots\}$ uses its dual $M' = \{\alpha, \beta, \gamma, \ldots\}$ of characters and runs as follows

$$f'(\alpha) = \sum_a f(a)\alpha(a), \quad f(a) = \sum_\alpha f'(\alpha)\alpha(a)^{-1}/n.$$

Here n is the order of M and sums run over M and M' respectively. To fit (1) and (2) into this format, note that w^{jk} is the value of the character $p \to \alpha(p) = w^{jk}$ of Z_n for $p = k$.

R. Let a be an element of M. Prove that there is a character β such that $\beta(a) \neq 1$ unless $a = 0$. (Hint. Use the decomposition of M.)

R. Show that $\sum \alpha(a) = 0$ when $a \neq 0$ and the sum runs over all characters. (Hint. Let $f(a)$ be the sum. Prove that $\beta(a)f(a) = 0$ for every character β and use the previous exercise.) Use this information for a direct proof of the inversion formulas above.

4.3 The finite Fourier transform and the quadratic reciprocity law

The finite Fourier transform leads to a simple proof of the quadratic reciprocity law
$$(p|q)(p|q) = (-1)^{(p-1)(q-1)/4}$$
where p and q are odd primes. Here $(n|p) = 1$ when n is a square mod q and -1 otherwise. This symbol has the property that

(1) $$\sum e(nj^2/p) = (n|p) \sum e(j^2/p)$$

when n is not divisible by p. Here $e(r) = e^{2\pi i r}$ and the sums, called Gaussian sums, run over Z_n. In fact, when n is a square mod p, then the sequence nj^2 and j^2 both run through the squares mod p so that the two sides are equal. When n is not a square mod p, the same sequences run through, respectively, the non-squares and the squares so that the sum on left plus the sum on the right equals $\sum e(j/p)$ for $j = 0, \ldots, p-1$ which vanishes. This proves (1) since $(n|p) = -1$ in this case.

The connection with with the finite Fourier transform T_p on Z_n ,
$$T_p u(j) = \sum e(jk/p) u(k)$$
with the matrix $(e(jk/p))$, stems from the fact that its trace, tr T_p is precisely

(2) $$\text{tr} T_p = \sum e(j^2/p)$$

Combining (1) and (2) yields
$$(p|q)(q|p) \text{tr} T_p \text{tr} T_q = \sum\sum e(qj^2/p + pk^2/q).$$

Here the sum on the right equals $\sum e((pj+qk)^2/pq)$ which is nothing but tr T_{pq}. Hence the quadratic reciprocity law follows if we can evalute the right side of

(3) $$(p|q)(q|p) = \text{tr} T_{pq} / \text{tr} T_p \text{tr} T_q.$$

THEOREM. *The trace of $\sqrt{n} T_n$, $n > 2$, equals $(1 + i^{-n})/(1 + i^{-1})$.*

Before the proof, let us deduce the quadratic reciprocity law from the theorem and (3). By (3), tr$\sqrt{n}T_n$ equals $1+i, 1, 0, i$ according as $n \equiv 0, 1, 2, 3 (4)$. Hence, for odd n, $\sqrt{n}\text{tr}T_n$ equals 1 and i according as $n \equiv 1$ or

4.3 Finite FT and quadratic reciprocity

$n \equiv 3 \mod 4$. Hence the right side of (3) is 1 unless both p and q are -1 in which case it equals -1.

To prove the theorem we shall compute $\mathrm{tr} T_n$ by developing the function

$$g(x) = \sum_{0}^{n-1} f(x+k), \quad f(x) = e(ix^2/n) = e^{2\pi i x^2/n}$$

in a Fourier series. This will give two expressions for $\mathrm{tr} T_n = g(0)$. The function $g(x)$ is periodic with period 1 and its Fourier coefficients are

$$a_m = \int_0^1 g(x)e(-imx)dx$$

Replacing g by f in the integral, we get

$$\int_0^1 e(-imx + ix^2/n)dx = e(-inm^2/4)\int_0^1 e(i(x - nm/2)^2)dx$$

so that, replacing x by $x + k$ and summing,

$$a_m = e(-inm^2/4)\int_0^n e(i(x - nm/2)^2)dx.$$

The factor in front of the integral is 1 when m is even and i^{-m} when m is odd and the integral equals

$$\int_{(nm-n)/2}^{(nm+n)/2} e(ix^2)dx.$$

The union of the intervals of integration is R when m runs over the odd integers and over the even integers. Hence

(4) $\qquad g(0) = (1 + i^{-n})\int e(ix^2/n)dx = (1 + i^{-n})(\pi n/(-2\pi i))^{1/2}$

which equals

$$(1 + i^{-n})\sqrt{n(i/2)} = \sqrt{n}(1 + i^{-n})/(1 + i^{-1}).$$

Since $g(0) = \mathrm{tr} T_n$, this finishes the proof modulo the computation of the integral in (4) and the fact that a continuously differentiable function with period 2π is the sum of its Fourier series.

4.4 The fast Fourier transform

The ordinary Fourier transform is a tool of supreme importance in analysis and has all kinds of applications, especially in electronics. Its importance stems from the fact that convolution appears naturally in processes where something that happens at a certain point in time depends linearly on what has happened before that time. Since the Fourier transform turns a convolution into a simpler operation, multiplication, it is very useful in analyzing such processes.

Numerically, the Fourier transform has to be implemented on a discrete grid and the computations should of course not take too long. From this point of view, the finite Fourier transform on a cyclic module presents certain disadvantages. In the first place, the roots of unity are complex numbers so that the Fourier transform leads to complex numbers. Secondly, to compute a Fourier transform with n arguments

$$f'(k) = \sum_{0}^{n-1} f(j) w^{jk}$$

seems to require $n(n-1)$ additions and the same number of multiplications. A smart bookkeeping trick, which is called the fast Fourier transform, FFT, makes it possible to do better when n is a composite number.

To shorten the exposition, a transform $S : f \to f'$ given by the formula

(1) $$f'(k) = \sum_{0}^{n-1} w^{jk} f(j)$$

will be called a Fourier transform of order n also when w is a primitive nth root of some complex number, not necessarily equal to 1.

First, let us observe that the formula (1) shows that computing the values of a Fourier transform of order n of a function f amounts to computing the values of the polynomial

(2) $$P(z) = f(0) + f(1)z + \cdots + f(n-1)z^{n-1}$$

when z runs through all the nth roots of unity. The fast Fourier transform is based on the following lemma which could be phrased briefly as: the computation of a Fourier transform of order $n = pq$ reduces to the computation of p Fourier transforms of order q.

LEMMA 1. *Suppose that $n = pq$ factors in natural numbers p and q and let t be a complex number. The values of the polynomial (2) as z runs through the nth roots of t are the same as the values of*

$$P_u(z) = \sum_{k=0}^{q-1} \left(\sum_{j=0}^{p-1} f(jq + k) u^j \right) z^k$$

4.4 The fast Fourier transform

as z runs through the qth roots of u and u through the pth roots of t.

PROOF: Let w be a primitive pqth root of t. Then we can split the sum (2) as

$$(3) \qquad \sum_{k=0}^{q-1}(\sum_{j=0}^{p-1} f(jq+k)w^{jq})w^k.$$

Putting $z = w$ and $u = w^q$, (3) follows.

Example
Let $n=4$ and $t=1$. With $u = 1, -1$, the two polynomials above are

$$P_1(z) = (f(0) + f(2)) + (f(1) + f(3))z$$

and

$$P_{-1}(z) = (f(0) - f(2)) + (f(1) - f(3))z.$$

For each of these polynomials we can apply lemma 1 again with $u = \pm 1$ and $u = \pm i$. Hence the four values of the transform of f are

$$(f(0) + f(2)) + (f(1) + f(3)), \quad (f(0) + f(2)) - (f(1) + f(3)),$$

$$(f(0) - f(2)) + (f(1) - f(2))i, \quad (f(0) - f(2)) - (f(1) - f(3))i.$$

The total number of additions is 4+4=8.

Cost estimates for FFT

The Fast Fourier transform, FFT, is a cover name for numerical schemes which split the computation of a finite Fourier transform in steps using Lemma 1. That the cost measured in terms of the number of additions and multiplications may be drastic will be seen below.

As remarked above, a straightforward computation of the Fourier transform (1) requires $n(n-1)$ additions and at most the same number of multiplications. On the other and, if $n = pq$ and Lemma 1 is used, $q(p-1)$ are required just to compute the coefficients of $P_u(z)$ and then $q(q-1)$ additions are required to compute all the values of P_u. Hence the total number of additions for the p functions P_u and hence for the computation of the Fourier transform of f is at most

$$(4) \qquad qp(p-1) + pq(q-1) = pq(p-1+q-1)$$

A count of the number of multiplications gives at most the same number. Since the last factor on the right of (4) is in general considerably less than

$pq - 1$, the use of Lemma 1 has greatly improved the economy of the computation of the Fourier transform of f. The theorem to follow uses the full power of Lemma 1.

The number of additions, subtractions included, used in some method of computation will be called its *additive* cost and the total number of additions and multiplications its *arithmetic* cost.

Example

The additive cost of a straightforward computation of a Fourier transform of order n is precisely $n(n-1)$, the arithmetic cost at most $2n(n-1)$.

THEOREM. *Let p and q be natural numbers. Suppose that $a(p)$ and $a(q)$ are the arithmetic costs of computing Fourier transforms of order p and q respectively. If the formula (3) is used to compute a Fourier transform of order pq, the corresponding arithmetic cost is*

$$(5) \qquad pa(q) + qa(p).$$

The same inequality holds when a denotes additive cost.

PROOF: There are p polynomials P_u of degree q to consider. Together they have pq coefficients which appear in groups of p such that all coefficients in one group are the values of a Fourier transform of order p. Hence the arithmetic cost of computing all coefficients is $qa(p)$. The arithmetic cost of computing the values of all polynomials P_u of degree q is also the cost of computing p Fourier transforms of order q. Hence (5) follows and the same reasoning for additive cost finishes the proof of the theorem.

Relative arithmetic cost. The theorem invites a new notion, namely the *relative* arithmetic cost or arithmetic cost per order of a Fourier transform, defined to be $c(m) = a(m)/m$. According to (5), the relative arithmetic cost of computing a Fourier transform of order pq is at most $c(p) + c(q)$ where $c(p)$ and $c(q)$ are the relative arithmetic costs for transforms of order p and q respectively. The same goes for the additive cost. Using these terms, the theorem has the following corollary.

COROLLARY. *Let $n = p^m$. Using Lemma 1 repeatedly, the relative arithmetic cost of computing a Fourier transform of order n is $mc(p)$ and hence at most $2m(p-1)$. The arithmetic cost itself is at most $2p^{m+1}m(p-1)$. In particular, when $p = 2$, the arithmetic cost is at most $2^{m+1}m$ and the additive cost at most half that amount.*

Note. These upper bounds are not the best possible ones (see for instance Auslander-Feig-Winograd (1984) and Winograd (1980)). Best bounds also depend on the class of algorithms used. For $p = 2$ it will be proved in

4.4 The fast Fourier transform

section 6.3 that the additive estimate is best possible for a certain class of algorithms which includes FFT. It will also be proved that FFT uses exactly $(n-1)2^n + 1$ multiplications by roots of unity not equal to 1 and that this is also best possible.

The case used in practice is $p = 2$. When computing a finite Fourier transform of order n, one is not bound by n being a power of 2. It is possible to choose a number k such that $2^k \le n < 2^{k+1}$ and, when $n > 2^k$, imbed functions with n arguments as functions of 2^{k+1} arguments with the last $2^{k+1} - n$ arguments put equal to zero. By the lemma above, it is then possible to compute the corresponding Fourier transform using at most $O(n \log n)$ additions and multiplications.

R. Let $n = \prod p^m$ be the prime decomposition of a natural number n, p running over the primes dividing dividing n and p^m is the corresponding primary number. Prove that the relative arithmetic cost of computing a Fourier transform of order n is at most $\sum mc(p)$, where $c(p)$ is the relative arithmetic cost for a transform of order p.

An algorithm for FFT of order 2^m

The 2^pth roots of unity will be parametrized by the formula

$$e(J) = \exp 2\pi i [J]/2^p$$

where $[J] = j_0 + \cdots + j_{p-1}2^{p-1}$ has $|J| = p$ binary digits. Then the two square roots of $e(J)$ are $e(J,j)$ with $j = 0,1$. Let

$$P(z) = \sum f([K])z^{[K]}, \quad |K| = m$$

be the polynomial associated to the Fourier transform of order 2^m with the coefficients numbered using m binary digits. We shall present an algorithm based on Lemma 1 which computes the Fourier transform of f using m generations of polynomials. Those of generation p are 2^p in number and have the degree $2^{m-p} - 1$. They will be denoted by

$$P(J,z) = \sum a(J;K)z^{[K]}$$

where $|J| = p$ and the sum runs over all K with $m - p$ digits.

FFT algorithm

Starting from $P(J,z) = P(z)$ when $|J| = 0$, construct the polynomial $P(J,j,z)$ from $P(J,z)$ and the binary digit j by the formula

$$a(J,j;L) = a(J;L,0) + e(J,j)a(J;L,1)$$

where $|J| + |K| = m - 1$. End: the polynomials with $|J| = m$.

This algorithm is justified by the following

LEMMA 2. *The polynomials $P(J,z)$ have the property that $p(J,z) = p(z)$ when z is an $(m-|J|)$th root of $e(J)$ and all values of $P(z)$ when z is an 2^mth root of unity is obtained in this way.*

Note. Apart from complicated notations, this is just Lemma 1.

PROOF: When $z^{2^{m-p}} = e(J)$, then $z^{2^{m-p-1}}$ is a square root of $e(J)$, $j = 0,1$. Then if $K = L, r$ with $r = 0, 1$, we have $[K] = [L] + 2^{m-p-1}r$, so that

$$z^{[K]} = z^{[L]}e(J,j)^r$$

and hence

$$P(J,z) = \sum a(J;L,r)z^{[L]}e(J,j)^r = \sum (a(J;L,0) + a(J;L,1)e(J,j))z^{[L]}.$$

An induction with respect to $|J|$ completes the proof.

R. Construct an analogous algorithm for Fourier transforms of order p^m.

Applications to convolutions and multiplication of polynomials

THEOREM. *Let f and g be complex functions from $0, \ldots, n-1$. Using FFT, the arithmetic cost of computing the convolution,*

$$(f \star g)(k) = \sum_{j=0}^{n-1} f(k-j)g(j)$$

or all the coefficients of the polynomial

$$(fg)(z) = f(z)g(z) = \sum_{j=0}^{2n-2}(\sum_{k=0}^{j} f(j-k)g(k))z^j$$

is $O(n \log n)$ in both cases.

R. Prove both assertions by using the formula

$$f = T^{-1}(TfTg).$$

The fast Fourier transform in modular arithmetic

In our cost estimates so far we have ignored the imprecisions of additions and multiplications which are inevitable when non-rational numbers are involved. In order to deal with the arithmetic of large integers it is often both necessary and convenient to use computation modulo some suitably chosen number. In this connection, a modular version of the fast Fourier transform has been used. In the modular version the complex numbers are replaced by Z_m and the primitive nth root of unity $e^{2\pi i/n}$ by an integer with the corresponding properties mod m. A suitable choice is the following

4.4 The fast Fourier transform

LEMMA. *Let n and w be positive powers of 2 and put $m = w^{n/2} + 1$. Then n has a multiplicative inverse mod m and*

(1) $$\sum_{0}^{n-1} w^{ip} \equiv 0 \mod m \text{ when } 1 \leq p \leq n-1.$$

Note. Since $w^{n/2} \equiv -1 \mod m$, w is a primitive nth root of unity in Z_m.

PROOF: Since m and n are obviously coprime, the first assertion follows. Let $n = 2^k$. The proof of the second depends on the identity

$$1 + a + a^2 + \cdots + a^{2^k - 1} = (1+a)(1+a^2)\ldots(1+a^{2^{k-1}}).$$

To prove this, it suffices to note that multiplication by $1 - a$ gives the same result $1 - a^{2^k}$ for both sides. When $a = 1$, the identity is obvious.

The identity above reduces the proof of (1) to a proof that

$$1 + w^{2^i} \equiv 0 \mod m$$

for some $i \geq 1$ and $< k$ depending on p. When p is odd, it suffices to choose $i = k-1$ for then $w^{2^{k-1}} = w^{n/2} \equiv -1 \mod m$. When p is not odd, $p = 2^t p'$, p' odd and $t < k$. Hence a choice of i realizing the desired congruence is always possible and this proves the lemma.

The Fourier transform Tf of order n of a function f with values in Z_m is now defined by

(2) $$(Tf)(j) = \sum_{k=0}^{n-1} w^{jk} f(k)$$

By the lemma above, this formula is inverted by

$$f(k) = n^{-1} \sum_{j=0}^{n-1} w^{-jk} (Tf)(j).$$

All calculations are of course performed in Z_m, i.e. modulo m.

THEOREM. *Let m and n be as in the Lemma. Using modular FFT, the computation mod m of a Fourier transform of order n (or its inverse) requires at most $O(n^2 \log n \log w)$ bit operations.*

PROOF: Since the algebra is the same as for the finite Fourier transform of order n, we can use FFT to compute the Fourier transform (2) or its inverse using at most $O(n \log n)$ additions and multiplications. It remains to estimate the bit cost of carrying out all these operations mod m. The integers mod m can be written with $b = (n/2) \log w + 1$ binary digits. Addition mod m then requires at most $O(b)$ binary operations and multiplication by powers of w also $O(b)$ binary operations. Hence the computation of the Fourier transform requires at most $O(n^2 \log n \log w)$ bit operations.

The Schönhage-Strassen algorithm for the multiplication of large integers

Under the bisection method, the bit cost of computing the product of two N-bit integers is $O(N^{1.8})$. A logarithmic bisection involving a step from N to $\sqrt{N}$ combined with a fast modular Fourier transform has led to a new algorithm, the Schönhage-Strassen algorithm, which computes the same product a the bit cost of $O(N \log N \log \log N)$. The interest of this algorithm, which has some delicate features, is mainly theoretical. The version presented below computes the product mod 2^N+1 which is sufficient since the order of the bit cost above is unchanged when when N is replaced by $2N$. For the same reason, it suffices to establish the algorithm when N is a power of 2.

The Schönhage-Strassen algorithm

Input: $N = 2^r$, r an integer > 0, two N-bit integers x and y.
Output: the product $z = xy \bmod 2^N + 1$.
1. If $r < 4$, compute z by some method.
2. If $r \geq 4$, set $b = n = 2^{r/2}$ when r is even and $n = 2^{(r-1)/2}, b = 2^{(r+1)/2}$, when r is odd. Set $m = 2^b + 1$ and $w = 2^{4b/n}$.
Note. By construction, $bn = N$ and $w = 2^4, 2^8$ according as r is even or odd. Since $w^{n/2} = 2^{2b}$, w is a primitive nth root of unity mod m.
3. Define $x(j) = y(j) = 0$ when $j < 0$ or $j > n-1$ and by

$$x = \sum_{j=0}^{n-1} x(j) 2^{jb}, \quad y = \sum_{j=0}^{j+n-1} y(j) 2^{jb}$$

otherwise. Define $z(j) = \sum_{j=0}^{n-1} x(j-k)y(k)$ and $t(j) = z(j) - z(j+n)$ when $j = 0, \ldots, n-1$.
Note. This defines $x(j)$ and $y(j)$ as b-bit integers. We also have

(1) $$xy \equiv \sum_{j=0}^{n-1} t(j) 2^{jb} \bmod 2^N + 1.$$

In fact, $2^{(n+j)b} z(j+n) \equiv -2^{jb} z(j+n) \bmod m$ since $2^{nb} = 2^N$. Also, $z(j) \leq j 2^{2b}$ and $z(n+j) \leq (n-j-1) 2^{2b}$ so that $(n-j-1) 2^{2b} \geq t(j) \geq -j 2^{2b}$.

4.4 The fast Fourier transform

The difference between the two bounds is $(n-1)2^{2b}$ and hence it suffices to know $t(j)$ modulo, for instance, $n(2^{2b}+1) = mn$ and, since the two factors are coprime, modulo m and n separately. This observation is the gist of the algorithm. The residue of $t(j)$ mod m will be computed using a modular Fourier transform and the algorithm itself with N replaced by $2b$. The residue modulo n, which is a much smaller number than m, is obtained by simpler means.

4. Define $u = 2^{2b/n}$ and $f(j) = u^j x(j)$, $g(j) = u^j y(j)$ for $j = 0, \ldots, n-1$. Let T be the Fourier transform mod m of order n with w as a primitive nth root of unity mod m. Put $F(j) = (Tf)(j), G(j) = (Tg)(j)$ for $j = 0, \ldots, n-1$. Define $h = T^{-1}(FG)$ where $(FG)(j) = F(j)G(j)$. Define

$$t'(j) \equiv u^{-j} h(j) \mod m.$$

Note. The powers of u stagger the b-bit integers $x(j)$ and $y(j)$ by inserting strings of zeros between them. Another essential effect of introducing the powers of u is that $t'(j) \equiv u^{-j} \mod m$. In fact, when $0 \leq j < n$, then $h(j) = u^j z(j) + u^{j+n} z(j+n) \equiv u^j t(j) \mod m$ since $u^n \equiv -1 \mod m$.

5. Compute F, G by the fast version of T, the products $F(j)G(j)$ of $2b$-bit numbers by the present algorithm, compute h by the fast version of T.

6. Define $x''(j), y''(j)$ to be the residues mod n of the integers $x(j)$ and $y(j)$ respectively. Putting $n = 2^t$, multiply the numbers

$$u = \sum_{j=0}^{n-1} x''(j) 2^{3tj}, v = \sum_{j=0}^{n-1} y''(j) 2^{3tj}$$

by the bisection method. The integers $z''(j) \equiv z(j) \mod n$ then appear as $2t$-bit sequences separated by sequences of t zeros. Compute $t''(j) \equiv z''(j) - z''(j+n)$.

7. Compute $t(j)$ exactly by the formula

$$t(j) \equiv m(t''(j) - t'(j)) + t'(j) \mod mn$$

and the condition that $t(j)$ be between $(n-1-j)2^{2b}$ and $(j+1)2^{2b}$.

Note. Since $b = n$ or $2n$, n and $m = 2^{2b} + 1$ are coprime and hence the right side above is $\equiv t(j) \mod mn$. By the note under step 3, $t(j)$ lies between the assigned boundaries.

8. Compute $xy \mod 2^N + 1$ using (1).

It is clear from the notes above that the algorithm achieves its purpose.

4 The finite Fourier transform

THEOREM. *The bit cost of the Schönhage-Strassen algorithm for the product of two N-bit integers is $O(N \log N \log \log N)$.*

PROOF: As remarked before, it suffices to verify the theorem when N is a power of 2 and the multiplication is done modulo $2^N + 1$. Let $M(N)$ be the actual bit cost in this case. By the previous theorem, the bit cost of step 5 above is $O(n^2 \log n + nM(2b))$. In fact, the first term is the bit cost of the three Fourier transforms and the last term majorizes the bit cost of computing the n terms $F(j)G(j)$ where each factor is at most a $2b$ bit number. Since the numbers u and v under step 6 have at most $3n \log n$ bits, the bit cost of multiplying them is at most $O((2n \log n)^{1.8}) = O(n^2)$. The other operations are shift which cost at most $O(N)$. Hence

$$M(N)/N \leq O(\log N) + (2bn/N)(M(2b)/2b).$$

To simplify this inequality, write $N = 2^r$ and $f(r) = M(2^r)/2^r$. By the definition of b, $2b$ equals $2^{h(r)}$ where $h(r) = r/2 + 1$ when r is even and $h(r) = r/2 + 3/2$ when r is odd. Hence we get

(2) $$f(r) \leq cr + 2f(h(r))$$

for some constant c and, for instance, $r \geq 4$. For $r < 4$, $f(r)$ is less than a fixed number a. Iterations of (2) give

(3) $$f(r) \leq \sum_{k=0}^{s} 2^k h^{(k)}(r) + 2^{s+1} f(h^{(s+1)}(r))$$

as long as $h^{(s)}(r) \geq 4$. Here $h^{(k)}$ is the kth iterate of h with $h^{(0)}(r) = r$. Since $r/2 + 1 \leq h(r) \leq r/2 + 3/2$, an easy calculation shows that

(4) $$2^{-k}r + 3 \geq h^{(k)}(r) \geq 2^{-k}r.$$

Since $h^{(k)}(r)$ decreases to zero as k increases, we can choose s such that

(5) $$h^{(s)}(r) \geq 4 \geq h^{(s+1)}(r).$$

According to (4), (5) implies that $2^{-s}r + 3 > 4 > 2^{-s}r$ which means that $2^s \leq r \leq 2^{s+3}$, i.e. $s \leq \log r \leq s + 3$. The estimate (4) shows that the sum of (3) is at most $(s+1)r + 3 \cdot 2^{s+1} = O(r \log r)$. By (5), the first term of the right side of (3) is at most $a2^{s+1}$. Hence $f(r) = O(r \log r)$ which proves the theorem. (Note the delicate balance between the factor 2 in (2) and the factor $1/2$ of r in $h(r)$.)

4.4 The fast Fourier transform

Literature

The main sources for this chapter have been Aho-Hopcroft-Ullman (1974) and Schönhage-Strassen (1971). The algorithm for the fast Fourier transform was invented by Cooley and Tukey (1965). The proof of the quadratic reciprocity theorem via the finite Fourier transfrom is taken from Auslander and Tolimieri (1979). This paper and that of Auslander, Feig and Winograd (1984) treat many aspects of the finite Fourier transform including other algorithms and cost estimates.

CHAPTER 5

Rings and fields

This chapter provides its reader with a quick view of the basic algebraic notions of ring and field illustrated with familiar examples. Its last section is devoted to the elements of linear algebra. The many easy R-exercises are supposed to make the reader familiar with the definitions and the terminology. The text is also a preparation for Chapter 6 on algebraic complexity theory.

5.1 Definitions and simple examples

A *ring* is a module $R = \{a, b, c, \dots\}$ with a multiplication, not necessarily commutative, and connected with the module operation by the distributive laws,
$$(a+b)c = ac + bc, \ c(a+b) = ca + ab.$$
When the multiplication is commutative, $ab = ba$ for all a, b in R, the ring is said to be *commutative*. A *subring* of a ring R is a non-empty part S of R such that $a - b$ and ab are in S when a and b are. It is then itself a ring, since the associative and distributive laws are automatic in S and, by a previous result, S is a module. The simplest of all rings is the ring Z of integers of which every mZ, m a fixed integer, is a subring. The rational numbers, the real numbers and the complex numbers are also rings. According to Chapter 1, all entire algebraic numbers constitute a ring and also all algebraic numbers. All of these are subrings of the ring of complex numbers.

These rings are all commutative. A very general model for rings, in general not commutative (see below under matrix rings) is the *endomorphism ring* $A = \mathrm{Hom}(M, M)$ of a module M, multiplication being defined by composition,
$$(fg)(x) = f(g(x)),$$
for all f, g in A and x in M.

R. Verify that fg is an endomorphism and that the associative and distributive laws hold.

R. Let a be an element of a ring A. Write na for $a + \cdots + a$ (n terms). Show that all finite sums
$$n_1 a + n_2 a^2 + \dots$$

5.1 Definitions and simple examples

with integral coefficients $n_1, \ldots$ form a subring of A and that it is the least subring containing a. Show that Aa, aA and all finite sums

$$b_1 a c_1 + b_2 a^2 c_2 + \ldots$$

with $b_1, c_1, \ldots$ in A form subrings of A.

Exercises

1. Let $R = \{a, b, c, \ldots\}$ be a ring and let $R' = \{a', b', c', \ldots\}$ be a copy of R as a set and isomorphic to R as a module. Define a product in R' by

$$a'b' = (ba)'.$$

Verify that R' is a ring. (R' is called the transpose of R.)

2. Let R be a ring and a, b two commuting elements of it. Show that the least subring of R containing a and b is the set of integral multiples of products of powers of a and b.

3. Which is the least subring of the real numbers containing i) $\sqrt{2}$ ii) 4 and $\sqrt{2}$ iii) $\sqrt{2}$ and $\sqrt{3}$?

Polynomial rings

The *polynomial ring* $R[x]$ over a ring R consists of all polynomials

$$f(x) = a_0 + a_1 x + a_2 x^2 + \ldots$$

(finite sums) with coefficients in R, x being an indeterminate commuting with everything in R, subject to the rule that a polynomial is zero if and only if all its coefficients are zero. If

$$g(x) = b_0 + b_1 x + \ldots$$

is another polynomial, the sum and difference $f \pm g$ and the product fg are defined by the familiar formulas

$$(f \pm g)(x) = a_0 \pm b_0 + (a_1 \pm b_1)x + \ldots, \quad (fg)(x) = a_0 b_0 + (a_0 b_1 + a_1 b_0)x + \ldots.$$

Also when R is not commutative it is easy to verify that $R[x]$, equipped with these operations, is a ring. It is a subring of the ring $R[[x]]$ of *formal power series* with coefficients in R. Its elements are the formal sums, possibly infinite,

$$f(x) = a_0 + a_1 x + \ldots.$$

Subtraction and multiplication are defined in the same way as for polynomials. Since a given coefficient of fg is a finite sum of products of coefficients of f and g, there are no difficulties with infinite sums of elements of R. The nth partial sum $f_n(x)$ of the formal power series is defined by

$$f_n(x) = a_0 + \cdots + a_n x^n$$

which we can also write as $f_n(x) \equiv f(x) \bmod x^{n+1}$. In this notation, the definition of an equality $f(x)g(x) = h(x)$ for formal power series is that $f(x)g(x) \equiv h(x) \bmod x^{n+1}$ for all integers $n \geq 0$.

Note. Formal power series with coefficients in Z_2 have applications to shift registers. See Chapter 8.

R. If $f(x)$, $g(x)$, $h(x)$ have coefficients a_i, b_j, c_k, prove that $(fg)(x)h(x)$ and $f(x)(gh)(x)$ have coefficients

$$d_l = \sum_{i+j+k=l} a_i b_j c_k.$$

This proves that multiplication of polynomials is associative. The proof of distributivity is similar, only simpler.

Direct sum and direct product

When $M = \{x, y, \ldots\}$ is a set and there is a ring R_x for every x in M, the *direct product*

$$\prod_{x \in M} R_x$$

consists of all functions $f(x)$ from M such that $f(x)$ is in R_x for all x. The ring operations are defined by

$$(f \pm g)(x) = f(x) \pm g(x), \ (fg)(x) = f(x)g(x).$$

The subring of $\prod R_x$ consisting of all functions $f(x)$ such that $f(x) \neq 0$ only for a finite number of x's is called the *direct sum* of the rings R_x. It is denoted by

$$\oplus_{x \in M} R_x.$$

R. Verify that the direct sum is a subring of the direct product.

When M has a finite number of elements, say $M = \{1, 2, \ldots, n\}$, the direct sum and the direct product are equal. We write them as

$$R_1 \oplus \cdots \oplus R_n.$$

R. When all R_k are equal to the real numbers R or the complex numbers C, the direct sum is a wellknown object. Which one?

Unit

A ring R may or may not have a *unit*, i.e., an element e such that $ae = ea = a$ for all a in R.

R. Show that $R[x]$ and $R[[x]]$ have units when R has a unit. Show that a direct product has a unit when all rings involved have units. Show that the direct sum $\oplus_{x \in M} R_x$ has a unit if and only if M is finite.

Exercises

1. Show that a ring cannot have more than one unit. (Hint. If there are two of them, consider their product.)

2. An element a of a ring R with a unit is said to have a left (right) inverse if there is a b in the ring such that $ba = e$ ($ab = e$). Show that if both exist, they must be equal.

Finite rings

The quotient modules $Z_m = Z/mZ, m > 0$, are examples of finite rings, the product of cosets $(a) = a + mZ$ being defined by

$$(a)(b) = (ab).$$

R. Prove that the right side is a function of the left side, i.e., that if a' and b' differ from a and b by multiples of m, then $a'b' - ab$ is a multiple of m. Prove associativity and distributivity.

Note. What has been done here is simply a pedantic verification that addition and multiplication mod m behaves just like ordinary addition and multiplication.

Characteristic

The integers m such that $ma = 0$ for all a in a ring R form a module Zn, $n = 0$ or > 0. The number n is called the *characteristic* of R. The characteristic of Z is 0, that of Z/mZ is m.

Matrix rings

Consider the set $M(n, R)$ of all $n \times n$ matrices $A = (a_{jk})$ with entries in a ring R. Addition and multiplication are defined by the usual rules

$$A + B = (a_{jk} + b_{jk}), \quad AB = \left(\sum_{i=1}^{n} a_{ji} b_{ik}\right).$$

R. Describe the zero and unit. Verify the associative and distributive laws. The set $M(n, R)$ is thus itself a ring.

R. Find two 2×2 matrices with integral entries which do not commute.

Matrices as endomorphisms

Consider the direct sum

$$X = \mathbb{Z}e_1 \oplus \cdots \oplus \mathbb{Z}e_n$$

of n infinite cyclic modules. When $x = c_1 e_1 + \cdots + c_n e_n$ is in X and $A = (a_{jk})$ is in $M(n, \mathbb{Z})$, put

$$Ax = \sum_{k=1}^{n} c_k A e_k, \quad A e_k = \sum_{i=1}^{n} a_{ik} e_i.$$

In this way, the matrix A is made into an endomorphism of X. In fact, by direct computation one finds that

$$A(x + y) = Ax + Ay, \quad A(Bx) = (AB)x$$

for all x, y in X and A, B in $M(n, \mathbb{Z})$. Hence we can think of the elements of $M(n, \mathbb{Z})$ as endomorphisms of the module X.

Exercise

Show that the following conditions on matrices $A = (a_{jk})$ give subrings of $M(n, R)$, R a commutative ring. i) Let $I_1, \ldots, I_t$ be a partition of $\{1, 2, \ldots, n\}$ and let $a_{jk} = 0$ unless j and k belong to the same set I_r. ii) $a_{jk} = 0$ when $j < k$. iii) $a_{jk} = 0$ when $j < k + p$, $p > 0$ fixed. Draw pictures under i) (for a simple partition) and under ii) and iii). Show that the subrings $R(p)$ defined under iii) have the property that $M(n, R)R(p) \subseteq R(p)$ and that $R(p)R(q) \subseteq R(p+q)$.

Invertible elements and zero divisors

An element a of a ring with unit whose multiplicative inverse a^{-1} exists is said to be *invertible*. When a is invertible and $ab = 0$ or $ba = 0$, we get $b = 0$ for $0 = a^{-1}ab = b$ and similarly in the other case. Products of invertible elements are invertible, for if a, b are invertible, then ab has the inverse $b^{-1}a^{-1}$. Typical non-invertible elements are the left (right) *zero divisors*, i.e., non-zero elements a for which $ab = 0$ for some b not equal to zero.

5.1 Definitions and simple examples

Examples
A matrix A in $M(n, \mathbb{Q})$ is invertible if and only if its determinant (denoted by det A) does not vanish (why?). When $A \in M(n, \mathbb{Z})$, A is invertible when det $A = 1$ or -1, but may not be invertible otherwise. When $R = \{a, b, \ldots\}$ is a ring, non-zero elements $(a, 0)$ and $(0, a)$ of the direct sum $R \oplus R$ are zero divisors unless R itself is zero.

R. Show that an element a of $\mathbb{Z}_m$ is invertible if and only if $(a, m) = 1$, and that the other non-zero elements are zero divisors.

R. Show that a formal power series

$$f(x) = 1 + a_1 x + a_2 x^2 + \ldots$$

with elements in a commutative ring R (with unit) is invertible. (Hint. Write $f(x) = 1 + xg(x)$. Show first that

$$h(x) = 1 - xg(x) + x^2 g(x)^2 + \ldots$$

is in fact a formal power series by checking that if the right side is written as a power series, then any coefficient is a *finite* sum of elements of R. Note that the degrees of the terms of a power $(xg(x))^k$ are at least k. Finally, to check that $fh = hf = 1$, check that $f(x)h(x) \equiv 1 \mod x^{n+1}$ for all $n \geq 0$.)

Exercises
1. Let R be a ring with a unit where every element equals its own square (Boolean ring). Show that R has characteristic 2 and is commutative. (Hint. Square $1 + x$ and $x + y$.)
2. Show that a ring without zero divisors containing a non-zero element e such that $e^2 = e$ must have e as a unit.
3. Let R be a finite ring with a unit. Show that an element is invertible if and only if it is not a zero divisor. (Hint. Look at the maps $R \to Ra$ and $R \to aR$.)

Division rings and fields

A ring in which every non-zero element is invertible is called a *division ring*. When commutative it is said to be a *field*. The classical fields are $\mathbb{Q}$, $\mathbb{R}$, $\mathbb{C}$ and the field of algebraic numbers.

R. Show that $\mathbb{Z}_m$ is a field if and only if m is a prime.

The simplest non-commutative division ring is the ring of *quaternions*, a real vector space generated by four elements $1, i, j, k$. A quaternion p is a linear combination

$$p = a_0 + a_1 i + a_2 j + a_3 k,$$

$a_k \in R$, where 1 is the unit and i, j, k have the following multiplication table

$$i^2 = j^2 = k^2 = -1, \ ij = -ji = k, \ jk = -kj = i, \ ki = -ik = j.$$

Notice the cyclic arrangement $ij = k, jk = i, ki = j$. The conjugate of a quaternion is defined by

$$\bar{p} = a_0 - a_1 i - a_2 j - a_3 k.$$

R. Use the rules above and the distributivity to show that

$$p\bar{p} = \bar{p}p = a_0^2 + \cdots + a_3^2,$$

and deduce from this that the quaternions form a division ring.

Note. We have left out the tedious verifications that the quaternions form a ring.

Exercises
1. Is the direct product of division rings a division ring?
2. Show that all $a + b\sqrt{2}$ with a and b rational numbers form a subfield of the real numbers.

Rings of quotients

The construction of fractions from the integers is a venerable invention of the human spirit. Without any difficulty it can be carried over to commutative rings.

Let R be a commutative ring and S a non-empty subset of R whose elements are not zero-divisors. Assume that S is multiplicatively closed, i.e., that st is in S when s and t are. Any such S can be used as denominators in a ring consisting of fractions a/s with a in R and s in S. We identify a/s with at/st for any t in S and add and multiply according to the usual rules

$$\frac{a}{s} + \frac{b}{t} = \frac{at + bs}{st}, \frac{a}{s}\frac{b}{t} = \frac{ab}{st}.$$

Simple but tedious verifications show that this construction produces a ring which we shall denote by R_S and describe as the *ring of quotients* of R with denominators in S.

R. Prove that R_S has a unit and that every s in S can be inverted in R_S.
R. What is R_S when $R = Z$ and $S = mZ \setminus 0$ with a fixed integer $m > 0$?

R. What is R_S when $R = \mathbb{Z}$ and S is generated by all primes except one, say p? Show that in this case p is the only prime in R_S being defined as it is for Z.

Integral domains

A commutative ring R is said to be an *integral domain* if it contains no zero divisors. Then we can take $S = R \setminus 0$ in the construction above.

R. Show that R_S is a field in this case.

The field R_S is called the *field of fractions*. Taking $R = \mathbb{Z}$ we get the rational numbers.

R. Prove that $R[x]$ is an integral domain when R is. (Hint. When a_n is not zero, we say that the polynomial

$$f(x) = a_0 + a_1 x + \cdots + a_n x^n$$

has *degree n*, and write $n = \deg f$. Prove that

$$\deg(fg) = \deg f + \deg g$$

for non-zero polynomials and go from there.)

Rational functions

Let F be a field. Then the ring $F[x]$ is an integral domain and the elements of its quotient field $F(x)$, i.e., quotients

$$\frac{f(x)}{g(x)}$$

where f and $g \neq 0$ are polynomials with coefficients in F, are called *rational functions*. If F is the field R or C, these are well-known from elementary analysis.

Mixed exercises for 5.1

1. Let $M = \mathbb{Z}x \neq 0$ be a module generated by a single element x. Prove that there is only one ring structure on M consistent with the requirement that $x^2 = mx$ for a fixed integer m. Prove that $x^j x^k = m^{j+k-1} x$ and that the ring so obtained is isomorphic to the ring $\mathbb{Z}m$.

2. Let $E(j,k)$ denote an $n \times n$ matrix which has 1 in the jth row and the kth column and zeros elsewhere. Check that they are multiplied by the rule that $E(j,k)E(p,q) = E(j,q)$ when $k = p$ and zero otherwise. Prove

that the entire ring $R = M(n, F)$, F a field, is generated by any non-zero matrix A in the sense that all sums of elements of RAR constitute all of R. (Hint. Write A as a linear combination of the $E(j,k)$ and start operating.)

3. Consider the module Q/Z, i.e., the rational numbers modulo the integers. Show that if a is in Q/Z, then a/n, naturally interpreted, is in Q/Z for all natural numbers $n \neq 0$. Deduce from this that $ab = 0$ for all a and b is the only way of making Q/Z into a ring respecting the module structure.

5.2 Modules over a ring. Ideals and morphisms

It is useful to let rings operate on modules the same way the integers operate on any module. A ring $R = \{a, b, c, \dots\}$ is said to *operate on the left* on a module $M = \{x, y, \dots\}$ if there is a function $(a, x) \to ax$ from $R \times M \to M$ which is associative and distributive in both factors,

$$a(bx) = (ab)x, \ (a+b)x = ax + bx, \ a(x+y) = ax + ay,$$

for all a, b in R and x, y in M. Operation on the right is defined in the same way: a function $(x, a) \to xa$ from $M \times R \to M$ such that

$$x(ab) = (xa)b, \ x(a+b) = xa + xb, \ (x+y)a = xa + ya.$$

When such operations are defined, M is said to be an *R-module* or a module over R. More precisely as the case may be, a left, right or two-sided R-module. When R is commutative, one does not distinguish between left and right modules.

R. Show that all $n \times 1$ matrices with elements in a ring R are natural left $M(n, R)$-modules. Is there a corresponding right module?

The ring R itself is of course an R-module, but there are plenty of others.

Example
For any ring R, the direct sum

$$S = R \oplus \cdots \oplus R$$

is a two-sided module. In fact, if $A = (a_1, \dots, a_n)$ is in S, put

$$aA = (aa_1, \dots, aa_n) \text{ and } Aa = (a_1 a, \dots, a_n a).$$

There are also R-modules properly contained in R. For a fixed $a \in R$, Ra is a left and aR a right R-module. All finite sums

$$b_1 a c_1 + b_2 a c_2 + \dots$$

5.2 Modules over a ring. Ideals and morphisms

with b_1, c_1 etc. arbitrary in R constitute a two-sided R-module.

Ideals

For an R-module contained in R there is another name, namely *ideal*. The word itself comes from an old divisibility theory for algebraic numbers based on 'ideal numbers' (Kummer, nineteenth century). As for R-modules there are left, right, and two-sided ideals. The set containing only the element 0 and R itself are the trivial ideals. All other ideals are said to be *proper*.

Any subset S of R generates a left ideal, viz. the set of finite sums

$$a_1 b_1 + a_2 b_2 + \ldots$$

with $b_1, b_2, \ldots$ in S and $a_1, a_2, \ldots$ in R. Ideals generated by a single element are said to be *principal*.

R. Show that every ideal is a subring and find an example where the converse does not hold.

R. Let $E(j, k)$ be the elementary matrices of $M(n, F)$, F a field. Show that the left ideal generated by $E(1, 1)$ consists of all linear combinations of $E(1,1), E(2,1), \ldots, E(n,1)$. Show that the entire ring considered as a left module over itself is the direct sum of n similar left ideals.

Note. An exercise at the end of the preceding section shows that 0 and the entire ring $M(n, F)$ are its only two-sided ideals. Rings with this property for ideals are said to be *simple*.

R. Let $R = \{a, b, c, \ldots\}$ be a ring and I a left ideal of R. Then, in particular, I is a submodule, and we may form the quotient module R/I. Show that the definition

$$b(a + I) = ba + I$$

makes R/I a left R-module (and not just a module).

R. Let R be a ring and I a *two-sided* ideal. Show that in this case the module R/I has a *ring structure* defined by

$$(a + I)(b + I) = ab + I.$$

The ring R/I is called the *quotient ring* of R over I.

Examples

All elements of the form $(0, s)$ form a two-sided ideal in the direct sum $R = T \oplus S$ of two rings. The elements of the quotient R/I are the cosets

(t, S) with $t \in T$. The submodule $m\mathbb{Z}$ of $\mathbb{Z}$ is also an ideal and the quotient $\mathbb{Z}/m\mathbb{Z}$ is nothing but the ring of integers mod m.

R. A proper ideal I in a commutative ring $R = \{a, b, c, \ldots\}$ with e as unit is said to be *maximal* if there is no ideal properly included between I and R. Show that the quotient R/I is a field if and only if I is maximal. (Hint. If a is in R but outside I, then $e = ab + c$ for some b and c in I.)

R. Show that a commutative ring $\neq 0$ is a field if and only if 0 and R are its only ideals.

R. Let R be the ring of continuous real functions $f(x)$ from a compact interval. Show that $f(x) = 0$ for x fixed defines a maximal ideal of R.

Exercise

Let R be a commutative ring with a unit. Show that an element of R is invertible if and only if it is outside every proper ideal.

Ring morphisms

A map $f : R \to R'$ from one ring $R = \{a, b, c, \ldots\}$ to another one $R' = \{a', b', c', \ldots\}$ is called a *homomorphism* or a *ring morphism* if

$$f(a - b) = f(a) - f(b), \ f(ab) = f(a)f(b)$$

for all a, b in R. If surjective and injective it is called an *isomorphism*. When $R = R'$ one uses the corresponding terms *endomorphism* and *automorphism*. The identical map $f(a) = a$ for all a is a trivial automorphism.

Examples

Let $R = S \oplus S$ be the direct sum of two copies of the same ring S. The maps $(a, b) \to (a, 0)$ and $(a, b) \to (0, b)$ are endomorphisms and the map $(a, b) \to (b, a)$ is an automorphism.

R. Let $R[x]$ be the ring of polynomials over a commutative ring R. Show that the maps $f(x) \to f(a)$ with a fixed in R are ring morphisms from $R[x]$ to R. (Hint. This follows directly from the definitions of sums, differences and products of polynomials.)

R. For a fixed integer a the map $x \to ax$ of the integers is a module morphism. Show that it is a ring morphism if and only if it is the identity.

R. Let c be a fixed invertible element of a ring $R = \{a, b, \ldots\}$. Show that

$$a \to cac^{-1}$$

is an automorphism of R.

5.2 Modules over a ring. Ideals and morphisms

Kernels and images

The *kernel* of a ring morphism $f : R \to R'$ is the set of elements of R which are mapped into the zero $0'$ of R'. It is denoted by ker f. The image of f, im f, is simply the subset $f(R)$ of R'.

R. Show that the kernel is a two-sided ideal of R and the image a subring of R'.

R. Let $F[x]$ be a polynomial ring over a field F. Describe the kernels and images of the maps $f(x) \to f(a)$.

We have now collected enough terminology and examples to state the morphism theorem for rings.

THEOREM. *When $f : R \to R'$ is a ring morphism, the image of f is isomorphic to the quotient $R/\ker f$ of R by the kernel of f.*

R. Prove the morphism theorem using your experience of the corresponding result for modules.

Exercise

Let $R = R_1 \oplus R_2$ be the direct sum of two rings. Put $f_1(a,b) = a$ and $f_2(a,b) = b$. Show that the maps $f_k : R \to R_k$ are surjective homomorphisms and that $R/\ker f_k$ is isomorphic to R_k.

Module morphisms

When M and N are two modules over a ring R, we say that a map $f : M \to N$ is an R-module morphism if $f(ax + by) = af(x) + bf(y)$ for all a,b in R and x,y in M (in case M and N are left R-modules). The kernel of f, ker f, is defined as the set of elements x of M such that $f(x) = 0$ and the image im f is $f(M)$. An R-module morphism f is said to be an *isomorphism* if it is bijective, an *endomorphism* if $M = N$ and an *automorphism* if it is an isomorphism from M to itself. A subset N of an R-module M is called a *submodule* if it is an R-module in itself, i.e., if $x - y$ and ax are in N when x,y are in N and a is in R.

R. Show that the kernel and the image are submodules of M and N respectively.

R. Show that an R-module morphism f is injective if and only if ker f is zero.

R. Let $N \subseteq M$ be modules over a ring R. Show that the quotient M/N is an R-module in a natural way.

R. Let M and N be two modules over a ring R and $f : M \to N$ an R-module morphism. Show that the quotient $M/\ker f$ is isomorphic to the image of f (the module morphism theorem).

R. Let M and N be two modules over a commutative ring R. Show that

$$(af + bg)(x) = af(x) + bg(x)$$

makes the set $\text{Hom}_R(M, N)$ of R-module morphisms from M to N into an R-module. Show that if $M = N$, then composition of morphisms gives a ring structure to $\text{Hom}_R(M, N)$.

R. Any commutative ring R is a module over itself. In particular, to every a in R there is an endomorphism $f_a : R \to R$ of R-modules defined by

$$f_a(x) = ax$$

for all x in R. Show that the map $a \to f_a$ from R to the endomorphism ring $\text{Hom}(R, R)$ of R is a ring morphism. Show that it reduces to an isomorphism when R has a unit.

Note. We saw earlier that $n \times n$ matrices with integer entries can be regarded as endomorphisms of the module $Ze_1 \oplus \cdots \oplus Ze_n$. In the same manner, $n \times n$ matrices with entries in a ring R can be regarded as endomorphisms of the module $R \oplus \cdots \oplus R$ with n terms (verify this).

Mixed exercises for 5.2

1. Are the matrix rings A and B with elements

$$\begin{pmatrix} a & b \\ 0 & c \end{pmatrix} \text{ and } \begin{pmatrix} 0 & a & b \\ 0 & 0 & c \\ 0 & 0 & 0 \end{pmatrix}$$

respectively, with a, b, c in a field isomorphic?

2. When d is an integer let $Z[\sqrt{d}]$ be the ring with elements $a + b\sqrt{d}$, a, b integers. Are the rings corresponding to $d = 2, 3$ isomorphic?

3. Show that the ideals given by i) $a = 0$, ii) $c = 0, b = da, d$ fix, iii) $a = b = 0$ are the only two-sided ideals different from zero and the ring itself in the ring of triangular matrices

$$\begin{pmatrix} a & b \\ 0 & c \end{pmatrix}$$

with elements in a field.

4. Let m and n be integers and coprime. Show that the ring Z_{mn} is isomorphic to the direct sum $Z_m \oplus Z_n$.

5. When I is an ideal in a commutative ring R, we let I^2 be the set of all (finite) sums $x_1 y_1 + x_2 y_2 + \ldots$ with x_i, y_i in I. Show that I^2 is an ideal. Let I be a minimal ideal in a commutative ring R, i.e., there are no ideal

properly contained between I and 0. Show that $I^2 = 0$ or else that I is a field.

6. An element a of a commutative ring R is said to be *nilpotent* if $a^n = 0$ for some natural number n. Show that the sum of two nilpotent elements is nilpotent. Also show that if b is in R and a is nilpotent, then ab is nilpotent. Conclude that the set of nilpotent elements form an ideal. This ideal is denoted by $N(R)$. Show that the quotient $R/N(R)$ does not have any nilpotent elements.

7. Let $f : R \to R'$ be a ring morphism. Show that the inverse image of an ideal of R', i.e., all the elements of R which are mapped by f into this ideal, form themselves an ideal.

8. Let $f : A \to B$ be a ring morphism. Show that f is injective or zero when A is a field.

9. Let R be a commutative ring with a unit and I, J two ideals such that $R = I + J$. Show that $R = I^2 + J^2$. (I^2 consists of all sums $\sum ab$ with a and b in I.)

10. An ideal P in a commutative ring R is said to be *prime* if a product cannot belong to P unless one of the factors does belong. Show that the quotient R/P is an integral domain if and only if P is prime.

11. Show that the polynomials $2x - 1$ and $x - 2$ do not generate the entire ring $Z[x]$.

12. Describe all ideals of the ring $Z \oplus Z$ and the corresponding quotients.

13. Describe the integers m for which Z_m contains nilpotent elements different from zero.

14. Let m be a natural number. Show that the ring Z_m does not contain a field unless m is prime. For another natural number n, let A be the set of elements x of Z_m for which $nx = 0$. Show that A is a ring with $k = m/(m,n)$ elements generated by the class of (m,n) in Z_m.

15. Let z and w be Gaussian integers and (z) and (w) the corresponding principal ideals in $Z[i]$. Show that $Z[i]/(zx)$ is isomorphic to the direct sum $Z[i]/(z) \oplus Z[i]/(w)$ when z and w are coprime.

16. Let $a + ib$ be a Gaussian integer. Show that every coset of the ideal $(a + ib)$ contains a unique Gaussian integer $x + iy$ in the rectangle $0 < x < m, 0 < y < (a^2 + b^2)/m$ where m is the greatest common divisor of a and b. This proves in particular that the quotient $Z[i]/(a + ib)$ has $a^2 + b^2$ elements. (Hint. Show that mod $(a + ib)$ one can simultanously reduce the real part of a Gaussian integer mod m and its imaginary part mod $(a^2 + b^2)/m$.)

17. A ring A with a unit is said to be Boolean if $x^2 = x$ for all $x \in A$. Show that every prime ideal in A is maximal.

5.3 Abstract linear algebra

Modules over a field are called *vector spaces*. Their theory is largely independent of the field and hence is the same as for the case when the field is the field of the real numbers. Although the reader is certainly familiar with this variant, the short exposition of the theory given here may be useful.

Let $M = \{x, y, \ldots\}$ be a vector space over a field $F = \{a, b, \ldots\}$. A finite or infinite subset $S = \{x, y, \ldots, u, \ldots\}$ of M generates subspace, i.e., a submodule, of M, namely its *linear span* $L(S)$ defined as the set of *linear combinations*

$$w = ax + by + \cdots + fu + \ldots$$

of $x, y, \ldots$ with coefficients $a, b, \ldots$ in F. The sum is supposed to be finite in the sense that at most a finite number of coefficients are different from zero. The elements of S are said to be *linearly independent* when a linear combination of them vanishes only when all its coefficients vanish. Otherwise they are *linearly dependent*. When the elements $x, y, \ldots, u, \ldots$ are linearly independent and $w \in L(\{x, y, \ldots\})$, the coefficients $a, b, \ldots$ are uniquely determined. For if $w = ax + by + \cdots = a'x + b'y + \ldots$, then $(a - a')x + (b - b')y + \cdots = 0$ and hence $a = a'$ etc.

Examples

The real numbers form a vector space over the rational numbers. The complex numbers form a vector space over the real numbers and also over the rational numbers. All real functions $f(t)$ from an interval T on the real axis form a vector space over the real numbers. The direct sum $F \oplus \cdots \oplus F$ of n copies of F is a vector space over F, which we shall denote by F^n.

The main results of linear algebra follow from

EXCHANGE LEMMA. *Let S be a subset of linearly independent elements of a vector space M and $x \neq 0$ an element of $L(S)$. Then there is an element y of S such that the elements of $T = (S \setminus y) \cup x$ are linearly independent and $L(S) = L(T)$.*

PROOF: By hypothesis, x is a linear combination

$$x = ay + bz + \ldots$$

of elements of S. Since $x \neq 0$ at least one of the coefficients does not vanish. Let it be a. Then

$$y = a^{-1}x - a^{-1}bz - \ldots$$

is in the linear span of $T = (S \setminus y) \cup x$. But then every element of $L(S)$ has the same property. The only way that the elements of T can be linearly

5.3 Abstract linear algebra

dependent is when x is a linear combination of the elements of $S \setminus y$. But this means that $a = 0$ in the expression above, which is a contradiction. The proof is finished.

Bases and dimension

A *basis* B of a vector space M is a collection of linearly independent elements such that $M = L(B)$.

Example

The elements $e_k = (0, \ldots, 1, 0, \ldots, 0)$ with 1 in the kth place form a basis of F^n.

LEMMA. *Let M be a vector space, S a subset of M such that $M = L(S)$, and assume that $T \subseteq S$ consists of linearly independent elements. Then there is a basis B of M such that $T \subseteq B \subseteq S$.*

PROOF: Note that if $x_1, x_2, \ldots$ are linearly independent and $x_0, x_1, x_2, \ldots$ are linearly dependent, then $x_0 \in L(\{x_1, x_2, \ldots\})$. Order the subsets of M consisting of linearly independent elements and containing T by inclusion. If $B_1 \subseteq B_2 \subseteq \ldots$ are subsets consisting of linearly independent elements and containing T, then their union $\cup B_k$ also consists of linearly independent elements. By Zorn's lemma, M has a maximal subset B of linearly independent elements such that $T \subseteq B$. If $x \in M \setminus B$, then, by the observation above, x is a linear combination of elements of B. Hence B is a basis.

Note. By the lemma, R has a basis as a vector space over Q. It has more than countably many elements (why?) and is called a Hamel basis. We note that it requires the axiom of choice (represented by Zorn's lemma) for its construction. Here we shall restrict ourselves to vector spaces which possess bases with a finite number of elements, the *finite-dimensional vector spaces*.

THEOREM. *In a finite dimensional vector space, all bases have the same number of elements.*

Notation. The number of elements in a basis of a vector space M is called the *dimension* of M and denoted by $\dim_F M$ or $\dim M$ if it is clear what the field is.

PROOF: Let B be a basis of M with n elements and suppose that B' is another basis. It is enough to prove that the number of elements of B' is $\leq n$. Suppose that B' has at least n elements. (Note that we do not assume that B' has a finite number of elements). By the exchange lemma, we can successively replace elements of B' by elements of B in such a way that all sets obtained consists of linearly independent elements and have the same linear span as B' (namely M). In the end we get a set $A = B \cup C$ where C is a subset of B' (possibly empty), such that the elements of A are linearly independent and $L(A) = M$. But by hypothesis, $L(B) = M$, wherefore the elements of C are linear combinations of the elements of B. But then C must be empty.

Coordinates

When $B = \{u_1, \ldots, u_n\}$ is a basis of a vector space M over a field F, every $x \in M$ is a unique linear combination

$$x = a_1 u_1 + \cdots + a_n u_n$$

of the basis elements with coefficients $a_1 = a_1(x), \ldots, a_n = a_n(x)$ in F. The numbers a_k are called the *coordinates* of x with respect to the basis B and the functions $x \to a_k(x)$ are called the *coordinate functions*. The coordinate functions a_k are examples of *linear functions*, i.e., morphisms f from one space M to another N. In fact, if x and y have the coordinates a_k and b_k respectively, then $ax + by$ has the coordinates $aa_k + bb_k$ (where a and b are elements of F).

R. Let $f_1, \ldots, f_n$ be coordinates on a vector space M. Prove that

$$x \to (f_1(x), \ldots, f_n(x))$$

is an isomorphism from M to F^n. In this sense every vector space of dimension n over a field F is essentially an F^n, but this statement does not do justice to the geometry of vector spaces.

Linear forms and dual spaces

Let M be finite dimensional vector space over a field F. Linear maps (i.e. morphisms) $f : M \to F$ are called *linear forms*. Example: coordinate functions with respect to a basis. All linear forms constitute themselves a vector space over F. We just define the linear combination $af + bg$ of two linear forms by

$$(af + bg)(x) = af(x) + bg(x)$$

for all x in M.

R. Verify that $af + bg$ is a linear function.

The vector space of all linear forms from a vector space M is called the *dual space* of M and will be denoted by M^*.

5.3 Abstract linear algebra

THEOREM. *If $B = \{u_1, \ldots, u_n\}$ is a basis of M, then the corresponding coordinate functions constitute a basis of M^*. In particular, $\dim M = \dim M^*$. If $f_1, \ldots, f_k$ are linearly independent in M^*, the equations*

$$f_1(x) = 0, \ldots, f_k(x) = 0$$

define a linear subspace of M of dimension $n - k$.

PROOF: The coordinate functions $f_1, \ldots, f_n$ are defined by the expansion

$$x = f_1(x)u_1 + \cdots + f_n(x)u_n$$

of x. Hence, if f is any linear form, then

$$f(x) = f_1(x)f(u_1) + \cdots + f_n(x)f(u_n)$$

which means that f is a linear combination of the coordinate functions. If

$$f(x) = c_1 f_1(x) + \cdots + c_n f_n(x)$$

is a linear combination of the coordinate functions which vanishes, it vanishes on every basis element u_k and hence $f(u_k) = c_k = 0$ for all k so that $f(x) = 0$ for all x. Hence the n coordinate functions constitute a basis of M^*.

To prove the second part we shall use induction. First, let $k = 1$. Then $f_1 = f \neq 0$ so that there is a u in M with $f(u) = 1$. Complete u to a basis $u_1 = u, u_2, \ldots, u_n$ of M. Then the elements

$$v_2 = u_2 - f(u_2)u, \ldots$$

are linearly independent and $f(v_k) = 0$ for all k and this is the desired result for $k = 1$. By induction we can assume that $f_2(x) = 0, \ldots, f_k(x) = 0$ defines a linear subspace of M of dimension $n - k + 1$. On this subspace $f_1(x)$ does not vanish and hence $f_1(x) = 0$ defines a subspace of dimension $n - k$ with a basis constructed as above. This proves the theorem.

R. Prove that any $n = \dim M$ linearly independent linear forms on M are the coordinate functions of some basis of M.

The following result is useful.

LEMMA. *Let N be a linear subspace of a linear space M and let $x \in M \setminus N$. Then there is a linear form which vanishes on N and equals 1 on x. If*

$f_1, \ldots, f_n$ are coordinate functions on M there is a subset $g_1, \ldots, g_k$ of them for which
$$x \to (g_1(x), \ldots, g_k(x))$$
is an isomorphism from N to F^k.

PROOF: If C is a basis of N, the set $C \cup x$ is linearly independent, and can be completed to a basis B of M. The coordinate which vanishes on all elements of B except x has the desired property. To prove the second part, note that $f_1, \ldots, f_n$ are linear forms also on N. Let $g_1, \ldots, g_k$ be a linearly independent subset of them with a maximal number of elements. Now any linear form $h(x)$ on N extends to a linear form on M and hence to a linear form on M and hence to a linear combination of the f_j. But then its restriction to N is a linear combination of the g_j. Hence the g_j form a basis for all linear forms on N. This finishes the proof.

Bilinear forms

Let M be a vector space over a field F. A *bilinear form* on M is a function $f(x, y)$ from $M \times M$ to F which is linear in each argument separately,

$$f(ax + by, z) = af(x, z) + bf(y, z), \ f(x, ay + bz) = af(x, y) + bf(x, z)$$

for all $a, b \in F$ and $x, y, x \in M$. Such a form is said to be *non-degenerate* if

$$f(x, y) = 0 \text{ for all } y \Rightarrow x = 0, \ f(x, y) = 0 \text{ for all } x \Rightarrow y = 0.$$

R. Prove that f is entirely determined by its values $f(u_j, u_k)$ on pairs of elements of a basis and that these values can be given arbitrarily. (The theory of systems of linear equations then proves that each of the conditions above implies the other.)

R. Let $f(u_j, u_k) \neq 0$ when $j = k$ and 0 otherwise. Prove that f is non-degenerate.

R. Via the bilinear form f, every x gives rise to a linear form $y \to f(x, y)$. Prove that if some elements of M are linearly independent so are the corresponding linear forms when f is non-degenerate.

THEOREM. *When f is a non-degenerate form and $u_1, \ldots, u_k$ are linearly independent elements of M, the equations*

$$f(u_1, y) = 0, \ldots, f(u_k, y) = 0$$

define a linear subspace of M of dimension $\dim M - k$.

Note. When M is a vector space and N a subspace, the number $\dim M - \dim N$ is called the *codimension* of N in M, and denoted by $\operatorname{codim}_M N$.

5.3 Abstract linear algebra

R. Prove the theorem via the induced linear forms.

Note. The theorem shows that if the bilinear form is non-degenerate, then the construction of orthogonal complements in geometry carries over to the abstract case. Every linear subspace N of M has a kind of complement $N^\perp$ defined by the equations $f(x,y) = 0$ for all $x \in N$. It is of course sufficient to let x run through a basis of N. The dimension of $N^\perp$ is the codimension of N and conversely.

Quotients

Let M be a vector space and N a subspace. As usual, we can form the quotient M/N consisting of the congruence classes $x + N$.

THEOREM. $\dim M/N = \dim M - \dim N$

PROOF: Let C be a basis of N and complete it to a basis $B = C \cup D$ of M. Then the image of D in M/N is a basis of M/N, for all elements of D are linearly independent modulo the elements of $N = L(C)$, and every element of M is a linear combination of elements of D modulo $N = L(C)$. This proves the theorem.

Linear maps

An important part of linear algebra is the study of morphisms, or *linear maps* between vector spaces. That $f : M \to N$ is a linear map from the space M to the space N means that

$$f(ax + by) = af(x) + bf(y)$$

for all $a, b \in F$ and $x, y \in M$.

R. Prove that a linear function is determined by its values on a basis of M and that these values can be given arbitrarily in N.

R. Prove that the image $\operatorname{im} f = f(M)$ of a linear map f is a subspace of N and that the kernel $\ker f$ consisting of the elements x of M such that $f(x) = 0$ is a subspace of M. Prove also that f is injective if and only if $\ker f = 0$.

R. When F is a field and M, N two vector spaces over F, the space $\operatorname{Hom}_F(M, N)$ is usually denoted by $L(M, N)$. Prove that if M and N are finite-dimensional, $\dim L(M, N) = \dim M \dim N$.

R. When $f : M \to N$ is a morphism of two modules over a commutative ring F, prove the *module morphism theorem*: $M/\ker f$ is isomorphic to $\operatorname{im} f$. (Hint: Copy the proof of the corresponding theorem for modules over Z.)

Most of the properties of linear maps between vector spaces are given by the following

MAIN THEOREM. *If $f : M \to N$ is a linear map from a finite-dimensional vector space M to another vector space N, then*

$$\dim(\ker f) + \dim(\operatorname{im} f) = \dim M.$$

PROOF: We just collect some results:

$$\dim M - \dim(\ker f) = \dim M/\ker f = \dim(\operatorname{im} f).$$

We can use this theorem to prove a result on systems of linear equations. Consider the quadratic system

$$f_1(x) = a_{11}x_1 + \cdots + a_{1n}x_n = b_1$$
$$\vdots$$
$$f_n(x) = a_{n1}x_1 + \cdots + a_{nn}x_n = b_n,$$

where the x_k are unknowns and the b_k are elements of the field F. The f_k define a linear function $f : F^n \to F^n$ by $f(x) = (f_1(x), \ldots, f_n(x))$. Obviously the system is solvable if and only if $(b_1, \ldots, b_n)$ is in im f. Since im $f = F^n$ if and only if ker $f = 0$ by the main theorem, the system is solvable for all right sides if and only if the *homogeneous* system $f_1(x) = 0, \ldots, f_n(x) = 0$ has the unique solution $x = 0$.

R. Prove that if the homogeneous system only has the solution $x = 0$, then the system has a *unique* solution for every right side.

R. Let $f_1(x) = b_1, \ldots, f_m(x) = b_m$, $x = (x_1, \ldots, x_n)$, be a system of linear equations, not necessarily quadratic. Prove that it is solvable if and only if

$$c_1 f_1(x) + \cdots + c_m f_m(x) = 0 \Rightarrow c_1 b_1 + \cdots + c_m b_m = 0$$

for all sequences $c_1, \ldots, c_m$, where the f_k are considered as functions from F^n to F.

Literature

In spite of its abstract form, the material of this chapter is old and of varied origin. It can be found in almost any textbook on algebra.

CHAPTER 6

Algebraic complexity theory

Complexity theory is one of the spin-offs of computer science. So far it is a chapter with some solved and plenty of unsolved problems. One exception is the theory of cost of computations where additions and subtractions are considered to cost nothing and the total cost is measured in terms of the number of multiplications and divisions necessary to achieve the algorithm. The essential tool is very simple: abstract linear algebra, but the results are far from trivial. They are the subject of the second section of this chapter, the first one contains some preliminary generalities on rings generated by indeterminates.

6.1 Polynomial rings in several variables

Polynomial rings in one or several variables are special cases of more general constructions defined below.

Let $X = \{x, y, \ldots\}$ be a monoid, i.e. a set X equipped with an associative multiplication $x, y \to xy$ and let $R = \{a, b, \ldots\}$ be a ring. Provided X has the property

(P) every element of X is the product of at most a finite number of other elements, a possible unit excepted,

we are going to construct a ring $M = R[[X]]$ whose elements are all functions $x \to a(x)$ from X to R. We shall write them as formal power series over X with coefficients in R,

(1) $$f = \sum a(x)x,$$

the sum running over X. If $g = \sum b(x)x$ is another element of M and c is in R, we define the difference $f - g$, the product cf and the product fg as follows

(2) $$f - g = \sum (a(x) - b(x))x$$

(3) $$cf = \sum (ca(x))x$$

(4) $$fg = \sum c(z)z, \quad c(z) = \sum a(x)b(y) \quad \text{for} \quad xy = z.$$

Note that (P) means that all sums here are finite. By the exercise below, $R[[X]]$ is a ring. When both f and g are polynomials in the sense that the sum (1) is finite, i.e. $a(x) \neq 0$ for at most a finite number of elements of X and analogously for g, then fg is defined also without the condition (P). It also follows that the set $R[X]$ of polynomials (over X with coefficients in R) is itself a ring.

R. Prove that $R[[X]]$ is a ring when (P) holds. (Hint. The set of functions from X to R is a natural left R-module. Prove the associativity and distributivity by noting that $fg = \sum\sum a(x)b(y)xy$. Prove that $(fg)h$ and $f(gh)$, $f(g+h)$ and $fg+fh$, $(g+h)f$ and $gf+hf$ have the same coefficients.)

R. Prove that the rings above are commutative when X is commutative.

Free monoids

A set S with elements $u, v, \ldots$ is said to generate a monoid X if every element of X is a finite product of elements of S and X is said to be *free* when two products are the same if and only if they have the same non-unit factors in the same order. When X is commutative, this definition has to be changed to: X is free if any two products of the generators are equal if and only if they contain the same non-unit elements the same number of times, but the order between them is arbitrary. It is clear that the property P holds in both cases. The generators of a free monoid are said to be *indeterminates*, commuting or not commuting according as the monoid is commuting or not. Monoids with a finite number of generators are said to be *finitely generated*.

Rings of polynomials over free monoids

In the sequel we shall restrict ourselves to rings $R[X]$ of polynomials over finitely generated free commutative monoids X. The ring R is also supposed to be commutative. If the generators of X are $x(1), \ldots, x(n)$, we shall also write $R[X] = R[x(1), \ldots, x(n)]$ and denote a polynomial in this ring by
$$f(x(1), \ldots, x(n)) = \sum a(u)u$$
where each u, called a monomial, is a product
$$u = x(1)^{k(1)} \ldots x(n)^{k(n)}$$
where $k(1), \ldots, k(n)$ are non-negative integers. When they all vanish, u is the unique unit element. Rings of this kind have the following important property.

6.1 Polynomial rings

HOMOMORPHISM THEOREM. *Let $R[X]$ be a polynomial ring over a commutative ring R with free generators $x(1),\ldots,x(n)$ and let $y(1),\ldots,y(n)$ be any elements of a commutative ring S containing R. Then the map*

$$H : f(x(1),\ldots,x(n)) \to f(y(1),\ldots,y(n))$$

is a ring homomorphism $R[X] \to S$.

Note. We can take $S = R$ or let S be a subring of $R[x]$ generated by some of its generators and map the rest of the generators to elements of R.

PROOF: Since two polynomials are the same if and only if their coefficients are the same, the map above is well defined. If $f = \sum f(u)u$ is the expansion of f in monomials, the image $H(f)$ of f is $\sum a(u)H(u)$. Hence $H(f-g) = H(f)-H(g), H(af) = aH(f)$ when a is in R and, since, by the distributivity in S,

$$\sum a(u)H(u) \sum b(u)H(u) = \sum c(t)H(t),$$

where $c(t) = \sum a(u)b(v)$ for $uv = t$, we have $H(f)H(g) = H(fg)$. This finishes the proof.

R. Why must the ring S be commutative for H to be defined? (Hint. Consider polynomials in two indeterminates.)

Degree

By definition, any monomial in a polynomial ring $R[X]$ is a product of generators of the monoid X. The number of times a generator appears as a factor is called its multiplicity in the monomial. The number of generators of the product, the unit excluded and multiplicity included, is called the *degree* of the monomial. It follows that $\deg(uv) = \deg u + \deg v$ when u and v are monomials. If we define the degree of a polynomial $f = \sum a(u)u$ as the highest degree of any monomial u wich appears with a non-zero coefficient in f, and set the degree of 0 (the zero polynomial) to $-\infty$, we have the following rules

$$\deg(f+g) \leq \max(\deg f, \deg g), \quad \deg(fg) = \deg f + \deg g$$

provided R does not have zero divisors, i.e. $a \neq 0, b \neq 0 \Rightarrow ab \neq 0$ for all elements of R. Special case: R is a field.

R. Verify these rules for the examples below and then prove them in general.

Examples

The degrees of the polynomials $2xy + 3x^2z + 13y + 3$ and $x + y + z + z^{13}$ of $Z[x, y, z]$ are 3 and 13 respectively.

Homogeneity

A polynomial whose non-zero terms have all the same degree m is said to be *homogeneous* of degree m. The set of these polynomials together with the zero polynomial form a submodule $R_m[X]$ of $R[x]$ and one has

$$R_m[X]R_n[X] \subset R_{m+n}[X].$$

When a polynomial $f = f(x)$ is homogeneous of degree m then $f(ax) = a^m f(x)$ for all a in R.

Ideals and quotients

From now on we shall consider polynomial rings $F[X]$ over a field F. A polynomial ring $F[X]$ has plenty of ideals, for instance any ideal generated by an arbitrary set of polynomials. A simple such example is the ideal $I(m)$ of polynomials $f = \sum a(u)u$ such that $a(u) = 0$ when $\deg u \leq$ a fixed number m.

R. Verify that I_m is an ideal.

The quotient $Q(m) = F[X]/I_m$ can be desribed as follows. $Q(0)$ is isomorphic to F. As a module, $Q(1)$ is isomorphic to the direct sum

$$F \oplus Fx(1) \oplus \cdots \oplus Fx(n)$$

where $x(1), \ldots, x(n)$ are the generators of X. Its ring structure is given by the rule that $x(j)x(k) = 0$ for all j and k. In the general case, computations mod $I(m)$ are performed by throwing away all terms of a polynomial of degree $\geq m$ before and after performing the ring operations.

Exercise

Multiply the polynomials $ax^2 + by + z$ and $xy^2 + z + xy$ in the ring $Z[x, y, z]$ mod $I(4)$. (Answer: $ax^2z + byz + bxy^2 + z^2 + xyz$).

Linear dependence

A number of polynomials $f_1(x), \ldots, f_n(x)$ of a polynomial ring $F[X]$ where F is a field generate a linear space $Ff_1(x) + \cdots + Ff_n(x)$. To decide whether they are linearly independent, i.e. if

$$a_1 f_1(x) + \cdots + a_n f_n(x) = 0$$

6.2 Multiplicative complexity

with coefficients in F implies that the coefficients vanish, it is often convenient to use the Homomorphism Theorem above. In general, it suffices to substitute for $x = (x(1), \ldots, x(n))$ or part of it different values in F which gives a system of linear equations for the coefficients.

Example

Are the polynomals $xy, xz, xy + x^2$ linearly independent when $F = Q$? Consider an equation $axy + bxz + c(xy + x^2) = 0$. With $y = 0$, we get $bxz + cx^2 = 0$. Putting here $x = 1$ and $z = 1, -1$ gives $b + c = 0, -b + c = 0$ so that $b = c = 0$ and hence also $a = 0$.

Note. Two non-zero polynomials whch are linear combinations of separate sets of monomials are of course linearly independent. Special case: two polynomials of different degrees of homogeneity.

Note. Whether or not two polynomials are linearly independent depends on the choice of the field F. The polynomials $xy + yz$ and $xy - yz$ are linearly independent when $F = Q$, but they are the same polynomial when $F = Z_2$.

6.2 Complexity with respect to multiplication

We now have more than enough preparation for the study of algebraic complexity theory. First we need a formal definition of an algorithm. Let F be a field and G a ring which is also a F-module.

Definition

An algorithm A consists of
1) a subset of B called the input to A,
2) a computation C which is a sequence

$$C = (f(1), \ldots, f(n))$$

of elements of G called the steps of A. The first steps, called the input steps, are the elements of B. Each following step is then either a linear function of the preceding steps (linear step) or a product of two such linear functions (product step).
3) an output D which is simply a part of C.

Note. That $f(k)$ is a linear function of the preceding steps means that it is a linear combination of them with coefficients in F plus an element of F.

The number of product steps of an algortihm is also referred to as its multiplicative complexity (cost). Two algorithms with the same input and output are said to be equivalent and an algorithm with a given input and minimal multiplicative complexity is said to be optimal. The following simple application of abstract linear algebra provides a lower bound for the multiplicative complexity of an algorithm with a given output. When H is any part of G, let $L(H)$ be the linear span of H with coefficients in F.

THEOREM. *Let A be an algorithm and let $M(A)$ be the number of product steps of C. Then*

$$\dim(L(D)+L(B))/L(B) \leq \dim L(C)/L(B) \leq M(A).$$

Note. The left side depends only on the input and output of the algorithm. This means that if equality holds at both places, the algorithm is optimal in the sense given above.

PROOF: Let $C(k)$ be the first k steps of C. When $f(k+1)$ is a linear step, $\dim L(C(k+1))/L(B)$ equals $\dim L(C(k))/L(B)$. Hence only the product steps increase $\dim L(C(k))/L(B)$. This proves the theorem.

Applications

In the applications of this general result, the elements of the input B are taken as indeterminates generating the ring G which may be commutative or not. When D has many elements, the lower bound given by the theorem may be the number of product steps of an optimal algorithm. Here are a few examples.

1) A product $(x+ty)(z+tu) = xz + t(xu+yz) + t^2yu$ where t is a power of 10, input $B = (x,y,z,u)$ and output $D = (xz, xu+yz, yu)$, i.e. the coefficients of the powers of t in the product. The field F is the rational numbers. Here $\dim(L(D)+L(B))/L(B)$ is 3 and there is an optimal algorithm as follows:

$$x,y,z,u,xz,yu,(x+z)(z+u),(x+z)(z+u)-xz-yu(=xu+yz).$$

As we have seen in section 2.1, repeated applications of this algorithm reduces somewhat the cost of multiplication of large integers compared to the cost of the traditional algorithm.

2) Products of square matrices. Let $x = (x(j,k))$ and $y = (y(j,k))$ be two $n \times n$ matrices and $z = (z(j,k))$ their product defined by

$$z(j,k) = \sum_{i=1}^{n} x(j,i)y(i,k).$$

Let A be an algorithm with input $B = x,y$ and output $D = z$. If the elements of x and y are taken as indeterminates, $\dim(L(D)+L(B))/L(B)$

is n^2. In fact, $L(D) \cap L(B) = 0$ by homogeneity and all elements $z(j,k)$ are linearly independent. To see this, put all elements of x equal to zero except $x(j,j)$ and all elements of y equal to zero except $y(j,k)$. Then $z(j,k) = x(j,j)y(j,k)$ is the only non-zero element of z. Hence the elements of z are linearly independent over any field F. It follows that any algorithm computing the product of two $n \times n$ matrices has at least n^2 product steps. This seems to be a bad lower bound since an algorithm which computes all the products of the formula above separately has n^3 product steps and it seems difficult to do any better. It was therefore somewhat of a sensation when Strassen (1969) found an algorithm which computes the product using less than n^3 product steps, at least asymptotically.

The basis is an algorithm which computes the product

$$\begin{pmatrix} a & b \\ c & d \end{pmatrix} \begin{pmatrix} x & y \\ z & u \end{pmatrix} = \begin{pmatrix} ax+bz & ay+bu \\ cx+dz & cy+du \end{pmatrix}$$

of two 2×2 matrices with 7 product steps, one less than the maximum 8. The input is $(a, \ldots, u)$, the output the four elements of the product. The seven products are the following ones

$$A = (a-d)(z+u), B = (a+d)(x+u), C = (a-c)(x+y),$$

$$D = (a+b)u, E = a(y-u), F = d(z-x), G = (c+d)x.$$

It is then a matter of straightforward verification to see that the four elements of the product matrix are

$$\begin{pmatrix} A+B-D-F & D+E \\ F+G & B-C+E-G \end{pmatrix}.$$

One interesting feature is that these identities hold also when the indeterminates $a, \ldots, u$ do not commute. This can be used to reduce the multiplication of two matrices of order $2k$ to seven multiplications of matrices of order k. A repetition of this procedure applied to two matrices whose order n is a power of 2, permits one to multiply them at the cost of $O(n^c)$ multiplications where $c = \log_2 7 = 2.81 \ldots$.

Note. Refinements of the procedure have given still better values of c. See Pan (1984) and, for the latest improvement $c = 2.376$, Coopersmith and Winograd (1987).

3) The product of two polynomials. Let

$$f(x) = a(0) + a(1)x + \cdots + a(n)x^n$$

and
$$g(x) = b(0) + b(1)x + \cdots + b(m)x^m$$
be two polynomials of degrees n and m and let
$$h(x) = c(0) + c(1)x + \cdots + c(m+n)x^{m+n}$$
be their product. The coefficients $c(k)$ of $h(x)$ are given by the formula
$$c(k) = \sum_{i+j=k} a(i)b(j)$$
where $a(-i) = a(n+i) = 0$ when $i < 0$ and correspondingly for the other polynomial. A count of the number of multiplications involved gives the number $(m+1)(n+1)$ for every coefficient of f meets every coefficient of g precisely once. On the other hand, if the coefficients of f and g are considered as indeterminates, the coefficients of h are $m+n+1$ in number and linearly independent over the linear space over some field F spanned by the coefficients of f and g. Hence any algorithm computing the coefficients of h require at least $n+m+1$ multiplications. An optimal algorithm is given by Lagrange's interpolation formula which runs as follows. Choose $p = m+n+1$ rational separate points $t(1), \ldots, t(p)$ (this may not be possible when the field F is finite) and let
$$Q_k(x) = \prod_{j \neq k}(x - t(j))/(t(k) - t(j))$$
so that $Q_k(t(j)) = 1$ when $j = k$ and zero otherwise. The difference
$$f(x)g(x) - \sum_1^{m+n+1} Q_k(x)f(t(k))g(t(k))$$
is then a polynomial of degree $m+n$ vanishing at $m+n+1$ separate points and hence zero. This permits us to recover the coefficients of $h(x)$ in a roundabout way from the $n+m+1$ products $f(t(k))g(t(k))$ of linear combinations of the coefficients of f and g by collecting all the coefficients of a given power of x. Since the number of operations in the field F is very large and introduces rounding errors when F is the rational numbers, the method may not be practicable in this case. This point illustrates one weakness of the axiomatic approach.

4) The interior product. Let x and y be vectors with n components $x(j)$ and $y(j)$ taken as indeterminates. Our theorem above applied to an algorithm A with input $B = (x, y)$ and output the interior product
$$z = x(1)y(1) + \cdots + x(n)y(n)$$

6.2 Multiplicative complexity

gives the trivial result that A must contain one product step. The correct result is n steps and this may be proved as follows.

Assume that the result holds for $< n$ components. Let

$$(L(y) + M)(L'(y) + M')$$

be the first product step involving y of an algorithm A computing z. Here $L(y)$ and $L'(y)$ are linear functions of y with coefficients in F not both zero and M and M' are polynomials in x. That the first product step involving y has this form is clear since the preceding steps can only give linear functions of the input and polynomials in x. If we renumber the elements of y suitably, we can assume that L or L' contain $y(n)$. Hence there are $n-1$ elements of F, $h(1), \ldots, h(n-1)$, and a polynomial $h(0)$ in x such that the product step vanishes when

$$y(n) = h(0) + h(1)y(1) + \cdots + h(n-1)y(n-1).$$

Next, let A' be the algorithm we get from A by substituting the above value of $y(n)$ into A and putting $x(n) = 0$. Then A' has one product step less than A and computes

$$(x(1) + h(1))y(1) + \cdots + (x(n-1) + h(n-1))y(n-1).$$

Taking $x(j) + h(j)$ as new indeterminates and applying the induction hypothesis, we conclude that A' has at least $n-1$ product steps. Hence A has at least n of them and this finishes the proof.

Note. By the same kind of reasoning one proves that any algorithm computing an expression

$$a(1)y(1) + \cdots + a(n)y(n)$$

where $y(1), \ldots, y(n)$ are indeterminates and $a(1), \ldots, a(n)$ are elements of a ring generated by other indeterminates $x(1), \ldots, x(m)$, has at least q product steps where q is the dimension of the linear space over the field F spanned by the elements $b(1), \ldots, b(n)$ where $b(k) = a(k)$ minus its constant term. The same result extends to several expressions of the form above if the coefficients $a(1)$ etc. are replaced by the corresponding columns (Winograd (1980), Laksov (1986)).

Quadratic and bilinear algorithms

Consider the polynomial ring $G = F[x]$ with $x = (x(1), \ldots, x(n))$ commuting indeterminates. When $P(x)$ is a polynomial, let P_m be the part of P of degree $\leq m$. Since $P \to P_m$ is a ring homomorphism, we have the following simple result.

LEMMA. *Suppose that the input to an algorithm A consists of field F and a set of indeterminates $x(1), \ldots, x(n)$ and that the output consists of a collection of polynomials $P(x)$ of degree at most m. Then A is equivalent to an algorithm A' whose steps are polynomials of degree at most m.*

PROOF: Just compute all products modulo polynomials of degree $\geq m$ in all steps of A.

When $m = 2$ this rsult can be sharpened.

THEOREM. *An algorithm A as in the lemma with an output of degree at most two is equivalent to an algorithm with at most the same number of product steps where all of them are products of two linear functions of the input indeterminates.*

PROOF: By the lemma we may assume that the degrees of the steps are at most two. Since reduction modulo linear steps can be made without multiplication, we may also assume that if a linear step has the form $f+g+h$ where f, g, h are homogeneous polynomials of degrees 0,1,2 respectively, then we may split it into three linear steps f, g, h. Now let $(f+g+h)(f'+g'+h')$ with f, g, h homogeneous of degree 0,1,2 and the same for f', g', h be the first product step where f and f' are not both zero. Then f, h, g, f', g', h' all occur as earlier steps. Hence, modulo linear combinations of earlier steps and modulo terms of degree < 2, the product equals gg' which is a product of linear functions. Hence, by successive modifications of A, we can find an equivalent algorithm with the same number of product steps where these are as indicated. This finishes the proof.

An algorithm A as above is said to be *bilinear* when the input indeterminates are grouped into two sets $y = (y(1), \ldots, y(p))$ and $z = (z(1), \ldots, z(q))$ and all product steps are products of a linear function of y by a linear function of z. The output of the algorithm is then bilinear, i.e. a collection of bilinear forms

$$b(x, y) = \sum_{j,k} b(j, k) y(j) z(k).$$

The condition that a bilinear output should be produced by a bilinear algorithm is a real restriction. There are outputs which require more product steps when produced by a bilinear algorithm than by a general one (see Laksov 1986).

The problem of the minimal number of product steps for a number of bilinear forms

$$f_i(x, y) = \sum_{j,k} a(i, j, k) x(j) y(k), \quad i = 1, \ldots, r$$

6.2 Multiplicative complexity

and bilinear algorithms has an interesting connection with a different mathematical problem. In any bilinear algorithm, every every f_i is obtained as a sum

$$\sum_{j,k} b(j,k) L_j(x) M_k(y)$$

of products of linear forms in x and y. Hence, with $z(1), \ldots, z(n)$ as new indeterminates, we have

(1) $$\sum_{i,j,k} a(i,j,k) z(i) x(j) y(k) = \sum_k N_k(z) L_k(y) M_k(z)$$

where $N_k(z) = \sum b(i,k) z(i)$. The right side of (1), say $f(x,y,z)$, is known as a trilinear form in the indeterminates x, y, z and the least number of linear forms in x, y and z for which f can be written as above is called the rank of f. If n is the multiplicative complexity of a bilinear algorithm with output $f_1, \ldots, f_r$, the formulas above shows that the rank of f is at most n. On the other hand, if we identify the coefficients of $z(1), \ldots, z(n)$ in an identity of the form above, we see that every f_i is a linear combination of n products of linear forms. Hence we have proved the following result,

THEOREM. *The minimal multiplicative complexity of a number of bilinear forms*

$$f_i(x,y) = \sum_{j,k} a(i,j,k) x(j) y(k)$$

is precisely the rank of the trilinear form

$$f(x,y,z) = \sum a(i,j,k) z(i) x(j) y(k).$$

Note. There are no general methods for computing the rank of an arbitrary trilinear form, but when there is just one bilinear form, say $\sum_{j,k} a(j,k) x(j) y(k)$, in the output of a bilinear algorithm, its minimal multiplicative complexity is just the rank of the matrix $a(j,k)$.

The symmetry of the definition of the rank of a trilinear form $f(x,y,z) = \sum_{i,j,k} a(i,j,k) z(i) x(j) y(k)$ implies that the multiplicative complexities of the three bilinear algorithms with the outputs

$$\sum_{j,k} a(i,j,k) x(j) y(k), \quad \sum_{i,k} a(i,j,k) z(i) y(k), \quad \sum_{i,j} a(i,j,k) z(i) x(j)$$

are the same.

6.3 Appendix. The fast Fourier transform is optimal

We are going to show that the fast Fourier transform is optimal among a certain restricted class of algorithms.

A class of algorithms

For reasons which will be clear below, the algorithms we shall consider will be called *additive*. They concern the computation of expressions of the form
$$a(0)c_0(z) + \cdots + a(n)c_n(z)$$
where the coefficients $a(0), \ldots, a(n)$ are given indeterminates, z generates a cyclic group C commuting with the coefficients and the $c_k(z)$ are polynomials in z with integral coefficients. The unit of C is identified with the integer 1 and the integer n is fixed. It is clear that all such expressions, conveniently called strings, form a $Z[z]$-module in a natural way.

An additive algorithm has the following properties,

(i) it computes a finite number of strings

(1) $$P_m(z) = a(0)z^{m(0)} + a(1)z^{m(1)} + \cdots + a(n)z^{m(n)}$$

where the $m(0), \ldots, m(n)$ are non-negative integers,
(ii) The algorithm consists of a finite sequence of strings
$$p_0(z), p_1(z), \ldots, p_j(z), \ldots$$
the first $n+1$ being the indeterminates $a(0), \ldots, a(n)$. The sequence finally contains all $P_m(z)$.
(iii) When $j > n+1$ strings are constructed, $p_j(z)$ is either the product of an earlier string by z or else the sum of two such strings.

When a string $p_j(z)$ is constructed using one or two others according to (iii), we say that these *precede* $p_j(z)$. It is clear that this defines a transitive relation 'precedes' and that only the strings which precede a $P_m(z)$ are necessary for the computation of it.

We can also conclude that any such string either does not contain a given indeterminate $a(i)$ or else contains $a(i)$ multiplied by some power of z. In fact, the algorithm only permits additions so that if another situation (multiplication by a general polynomial) occurs, it must also occur in some $P_m(z)$ which is impossible. (Note that this argument holds also when $z^m = 1$ for some $m > 0$.) Observe also that if the algorithm also permits subtractions, the polynomials which precede some polynomial to be computed may have the general form (1). Then the proof below is no

6.3 The fast Fourier transform is optimal

longer valid and FFT may not be optimal. This is in fact also the case. Winograd (1980) has shown that this kind of algorithms require at least $2^n - 2$ multiplications, a bound which has been obtained for at least $n = 3$. Nevertheless, this section may be of some interest.

Example

When just one string (polynomial), $P = a(0) + a(1)z + \cdots + a(n)z^n$ with $n > 0$ is to be computed, it is obvious that at least n additions are required. We shall see that at least the same number of multiplications are required. In fact, this is true for $n = 1$ and since any algorithm which computes P also computes P when $a(n) = 0$, it must, by induction, have $n - 1$ multiplications. But it also computes P when $a(n)$ is arbitrary and this involves at least one additional multiplication. Note that Horner's algorithm realizes the optimal number of additions and multiplications at the same time.

The FFT algorithm uses only the steps of an additive algorithm in the sense above. We shall see below that it is an optimal one.

LEMMA 1. *All polynomials of an algorithm above preceding $P_m(z)$ have the form*

$$(1) \qquad z^p(a(i)z^{m(i)} + a(k)z^{m(k)} + \ldots)$$

where $p \geq 0$ and $i, k, \ldots$ are all separate.

PROOF: We know that they have the form above with $a(i), a(k), \ldots$ separate but multiplied by some powers of z. Consider such a polynomial,

$$Q(z) = a(i)z^{t(i)} + a(k)z^{t(j)} + \cdots.$$

Then the polynomials which it precedes can only have the form $z^p Q(z) + R(z)$ where $R(z)$ does not contain the coefficients of $Q(z)$. This applies in particular to $P_m(z)$ itself and hence

$$p + t(i) = m(i), \ p + t(k) = m(k), \ldots.$$

This proves the lemma.

We remarked above that when an additive algorithm computes just one ordinary polynomial of degree n, it must necessarily involve at least $n - 1$ additions. If the values of more than one polynomial are computed, the number of additions may be less than $2(n - 1)$ if polynomials of the form (1) can be used for both of them. This happens very often in the fast Fourier transform and is the basis of its economy.

Application to the finite Fourier transform

Consider now the Fourier transform of order 2^n,

(2) $\quad F(u) = f(0) + f(1)u + \cdots + f(2)u^2 + \cdots + f(2^n - 1)u^{2^n - 1},$

where u runs through all 2^nth roots of unity. All these roots form a cyclic multiplicative group of order 2^n. Define the order of a root u of unity to be the least number $m > 0$ for which $u^m = 1$. By elementary theory, all orders of the 2^nth roots of unity are powers of 2. Those of order 2^n are said to be primitive. All primitive roots are odd powers of any one of them, the others are even powers.

We are going to find the minimal arithmetic cost of the Fourier transform when additive algorithms are used. This means in particular that the values $f(0), f(1), \ldots$ of f are considered to be indeterminates. It also means that the numerical 2^nth roots of unity are now replaced by the elements of a cyclic group C of order 2^n generated by z where z^{2^n} is identified with 1 but $z^{2^{n-1}}$ is not identified with -1.

Our polynomial algorithms compute all the values of F through a succession of expressions of the form (1). To have a convenient name, we shall call them substrings. The following lemma is important.

LEMMA 2. *When u is a primitive 2^nth root of unity, no substring of $F(u)$ is a substring of $F(v)$ when v is not a primitive 2^nth root of unity.*

PROOF: Suppose that a substring for $F(u)$ times some power of u is also a substring for $F(v)v$,

$$u^p(f(i)u^i + f(k)u^k + \ldots) = u^q(f(i)v^i + f(k)v^k + \ldots)$$

for some p, q and separate $i, k, \ldots$ and all $f(i), f(k), \ldots$. (Note that the form of the substrings follows from Lemma 1.) Since the $f(i), f(k), \ldots$ are arbitrary, this means that

$$u^{p+i} = u^q v^i, \quad u^{p+k} = u^q v^k$$

and hence that
$$(v/u)^{k-i} = 1.$$

Now, v being of order $< 2^n$ is some power of u,

$$v = u^c, \quad c = 2^t b, \quad b \text{ odd},$$

where $t > 0$ so that v/u equals u raised to some odd power. But then (2) shows that $i - k \equiv 0 \mod 2^n$. But this is impossible since $0 \leq i < k < 2^n$ and the proof is finished.

6.3 The fast Fourier transform is optimal

As in section 4.4, let us now define the additive and multiplicative costs of an additive algorithm to be, respectively, the number of additions and multiplications it uses. The sum of these numbers will be called its arithmetic cost. The following theorem completes the Corollary of section 4.4. and proves that the fast Fourier transform is optimal in a precise sense.

THEOREM. *The additive cost of an additive algorithm which computes a Fourier transform of order 2^n is at least $n2^n$. Its multiplicative cost is at least $(n-1)2^n + 1$.*

Note. According to section 4.4, the FFT algorithm realizes the minimal additive cost and it will be proved below that it also realizes the minimal multiplicative cost counted as the number of multiplications by roots of unity not equal to 1.

PROOF: Let us call the values $F(v)$ of the Fourier transform primitive when v is a primitive 2^nth root of unity. By Lemma 2, no subsum used to compute these occur in the computation of the other values. Hence the additive and multiplicative costs for an algorithm is the sum of these costs for two groups. Let $a(n)$ and $m(n)$ be the minimal additive and multiplicative costs for Fourier transforms of order 2^n and let $a_p(n)$ and $m_p(n)$ be the corresponding costs for the primitive roots.

Any algorithm which computes the $F(u)$ for u primitive also computes the same string where the second half $a(2^{n-1}),\ldots$ of the indeterminates are put equal to zero. Hence $a_p(n) \geq a(n-1)$. But this algorithm must also have at least one addition for every $F(v)$ bearing on the second half of the indeterminates. Hence

$$a_p(n) \geq a(n-1) + 2^{n-1}.$$

The same argument works for the strings $F(v)$ with v not primitive and hence the right side above majorizes also $a(n) - a_p(n)$. Adding the two, we get the recurrence

$$a(n) \geq 2a(n-1) + 2^n, \quad a(1) = 2$$

whose solution is $a(n) \geq n2^n$. The same arguments works for the multiplicative costs with the exception that no multiplications are needed in $F(v)$ for the second half of the indeterminates when v is no primitive. This gives the recurrence

$$m(n) \geq 2m(n-1) + 2^{n-1}, \quad m(1) = 1$$

whose solution is $m(n) \geq (n-1)2^n + 1$. This proves the theorem.

An inspection of the FFT algorithm of section 4.4 shows that the number of multiplications by roots of unity except 1 that it uses satisfies the same recurrence as $m(n)$ but with equality. Hence it is optimal with respect to multiplication and addition.

Literature

The main sources for this chapter are Winograd (1980) and Laksov (1986). Winograd's book contains a lot of material not covered here, for instance algorithms for finite Fourier transforms of prime order. The references for the latest results on the cost of matrix multiplication is Pan (1984) and Coppersmith-Winograd (1987). The theorem of section 6.3 seems to be new.

CHAPTER 7

Polynomial rings, algebraic fields, finite fields

The material of this chapter, divisibility in polynomial rings, algebraic numbers over a field and finite fields are substantial pieces of mathematics, over 150 years old. Some of the results have found applications in computer science.

7.1 Divisibility in a polynomial ring

When R is a commutative ring, let $R[x]$ be the ring of polynomials

$$f(x) = a_0 + a_1 x + a_2 x^2 + \cdots + a_n x^n$$

with coefficients in R. We shall see that most of the familiar properties of R[x] and C[x], the rings of polynomials with real and complex coefficients, carry over to the general case of an arbitrary field k, for instance $k = Z_p$, p a prime. We shall also deal with the ring Z[x] of polynomials with integer coefficients.

In the study of divisibility in polynomial rings, it is often necessary to compare polynomial rings over different integral domains or fields. It is therefore convenient to start with an

EXTENSION LEMMA. *Any homomorphism $\varphi : R \to R'$ from a commutative ring R to another one R', extends naturally to a homomorphism $R[x] \to R'[x]$ of the corresponding polynomial rings defined by*

$$\varphi(a_0 + a_1 x + a_2 x^2 + \ldots) = \varphi(a_0) + \varphi(a_1)x + \varphi(a_2)x^2 + \ldots.$$

When φ is an isomorphism, its extension has the same property.

Example
When $R = Z$, we have for instance

$$(3x + 2)(x^2 + 2x + 1) = 3x^3 + 8x^2 + 7x + 2.$$

Under the map $a \to a \bmod 2$ from Z to Z_2, this reduces to

$$x(x^2 + 1) \equiv x^3 + x \quad \bmod 2$$

(i.e. equality in the ring Z_2), which is also a true formula.

PROOF: When $f(x) = a_0 + a_1 x + \ldots$ is a polynomial in $R[x]$, put for simplicity $\varphi f(x) = \varphi(a_0) + \varphi(a_1)x + \ldots$. We have to show that $\varphi(f-g)(x) = \varphi f(x) - \varphi g(x)$ and $\varphi(fg)(x) = \varphi f(x)\varphi g(x)$ when

$$f(x) = \sum_{j=0}^{n} a_j x^j \text{ and } g(x) = \sum_{k=0}^{m} b_k x^k$$

are polynomials in $R[x]$. The first property is clear since

$$\varphi(f-g)(x) = \sum \varphi(a_j - b_j)x^j = \sum (\varphi(a_j) - \varphi(b_j))x^j,$$

where the right side is $\varphi f(x) - \varphi g(x)$. The coefficients of the product $f(x)g(x) = \sum_{i=0}^{m+n} c_i x^i$ are given by

$$c_i = \sum_{j+k=i} a_j b_k.$$

The coefficients d_i of the product $\varphi f(x)\varphi g(x)$ are obtained in the same way from the coefficients $\varphi(a_j)$ and $\varphi(b_k)$ of $\varphi f(x)$ and $\varphi g(x)$. Since

$$\varphi(c_i) = \sum_{j+k=i} \varphi(a_j)\varphi(b_k),$$

we have $d_i = \varphi(c_i)$ and hence $\varphi(f(x)g(x)) = \varphi(f(x))\varphi(g(x))$ and this finishes the proof since the last statement of the lemma is now obvious.

The binomial theorem in characteristic m

The binomial formula

$$(1+x)^n = \sum_{j=0}^{n} \binom{n}{j} x^j$$

and the corresponding formula

$$(1-x)^{-n} = \sum_{j \geq 0} \binom{n+j-1}{n-1} x^j$$

are true also in arbitrary characteristic. In fact, the coefficients are integers, and the formula in characteristic m is obtained simply by extending the homomorphism $Z \to Z_m$ to polynomial rings and formal power series.

R. Let $m = p$ be a prime. Show that $(1+x)^p \equiv 1 + x^p$ mod p and, more generally, that $g(x^p) \equiv g(x)^p$ mod p when $g(x)$ is a polynomial with integral coefficients.

Degree

Let us now restrict ourselves to polynomial rings over integral domains (i.e., commutative rings without zero divisors) and, as a special case, polynomial rings over fields. Let

$$f(x) = a_0 + a_1 x + \cdots + a_n x^n$$

be a polynomial with coefficients in an integral domain R. When $a_n \neq 0$, the last term on the right side is called the *leading term* of the polynomial, and a_n the *leading coefficient*.

We say that the *degree* of f, deg f, is n if $a_n \neq 0$ in the formula above. The degree of a constant polynomial $\neq 0$ is zero. The degree of the zero polynomial is not defined (although in some cases it is convenient to let its degree be $-\infty$).

Precisely as in the classical case one has

$$\deg(fg) = \deg f + \deg g$$

if f and g are non-zero. This equality expresses the fact that if ax^n and bx^m are the leading terms of f and g respectively, then abx^{n+m} is the leading term of fg and that $a, b \neq 0$ implies that $ab \neq 0$, since we are working over an integral domain. It follows in particular that $f(x)g(x) \neq 0$ when $f(x), g(x) \neq 0$, i.e., that $R[x]$ is an integral domain, too.

Divisibility and primes in a polynomial ring

The theory of divisibility in a polynomial ring $k[x]$ over a field k is analogous to the same theory for integers. When the word polynomial is used below, it refers to an element of $k[x]$ and the word *unit* refers to any non-zero element of k. Polynomials with leading coefficient 1 are called *monic*. It is clear that $h(x)$ is monic and $h(x) = f(x)g(x)$, then $f(x)$ and $g(x)$ can be chosen to be monic by multiplication by units.

A non-zero polynomial $f(x)$ is called a divisor or factor of a polynomial $h(x)$ when there is a third polynomial $g(x)$ such that $h(x) = f(x)g(x)$. Under the same circumstances, $f(x)$ is said to divide $h(x)$ and $h(x)$ is said to be a multiple of $f(x)$ (and of $g(x)$). Every polynomial has *trivial* divisors, namely units and unit multiples of itself.

A polynomial with only trivial divisors is said to be *prime* or *irreducible*. Powers of a prime polynomial are called *primary* and two polynomials whose only common divisors are units, for instance two different monic prime polynomials, are said to be *coprime*.

R. Using the properties of the degree, show that every prime polynomial has positive degree and that those of degree 1 are prime.

The theory of divisibility in a polynomial ring over a field k depends on the fact that the division algorithm for real and complex polynomials extends to polynomials with coefficients in a field: if f and g are polynomials and g does not vanish, there exist polynomials q and r such that $f = qg + r$ and $\deg r < \deg g$ or else $r = 0$.

R. A reader not familiar with the division algorithm can prove it by repeated applications of the following step: given a polynomial $f(x)$ and a polynomial $g(x)$ such that $\deg g < \deg f$, there exists a polynomial $s(x)$ such that $\deg(f - sg) < \deg f$. To see that s exists, let $f(x) = ax^n + \ldots, g(x) = bx^m + \ldots$ and put $s(x) = \frac{a}{b}x^{n-m}$. It is clear from this why k has to be a field for the algorithm to work.

R. Using the algorithm prove the *factor theorem*: a polynomial $f(x)$ is divisible by $x - a$, a in k, if and only if $f(a) = 0$.

R. Using the division algorithm, show that every ideal $I \neq 0$ in $k[x]$ is principal, i.e., has the form

$$I = k[x]f(x)$$

for some polynomial $f(x)$, unique apart from a constant factor, and called the *generator* of I. (Hint: There is an $f(x) \neq 0$ in I of lowest degree. Prove that the division algorithm gives a contradiction unless every element of I is a multiple of $f(x)$.) Prove that the generator $h(x)$ of the ideal $k[x]f(x) + k[x]g(x)$ is the greatest common divisor of $f(x)$ and $g(x)$.

Note. The generator of a non-zero ideal is uniquely determined if we require it to be monic.

R. Prove that if a prime polynomial divides a product, it divides at least one of the factors. (Hint: Suppose that $f(x)$ is prime and divides a product $g(x)h(x)$ but not $g(x)$. Then $f(x)$ and $g(x)$ cannot have a common non-trivial factor and hence, by the preceding exercise, there are polynomials $a(x)$ and $b(x)$ such that $1 = a(x)f(x) + b(x)g(x)$. Multiplying this identity by $h(x)$ shows that $f(x)$ divides $h(x)$.)

DIVISIBILITY THEOREM. *Every nonvanishing polynomial of positive degree in $k[x]$, k a field, is a product of primary polynomials, unique apart from the order and constant factors.*

Note. Restriction to monic polynomials makes the product unique up to order.

7.1 Divisibility in a polynomial ring

PROOF: The properties of the degree shows that every polynomial is a product of prime polynomials and hence also a product of primary polynomials belonging to prime polynomials which are pairwise coprime. Now, by an exercise above, a prime polynomial dividing $f(x)$ has to divide precisely one primary factor. Hence a primary factor of $f(x)$ is the highest power of the corresponding prime which divides $f(x)$. This proves that the primary factors are unique apart from their order and multiplications by units and finishes the proof.

Examples

When k is the field of complex numbers, the fundamental theorem of algebra tells us that every polynomial is the product (apart from constants $\neq 0$) of linear factors $x - a$, where a is a complex number. In particular these are all prime polynomials of the ring of complex polynomials.

This result does not apply to real polynomials for the simple reason that the complex prime polynomials $x - a$ may not have real coefficients.

R. Prove, using the previous result and the fact that conjugation of the coefficients preserves a polynomial with real coefficients, that the prime polynomials of R[x] are the polynomials of degree 1 and the polynomials of degree 2 with no real zeros, the most well-known of this kind being $x^2 + 1$.

Example

There is only one prime polynomial of degree two in $Z_2[x]$, namely $x^2 + x + 1$. In fact, one finds that the three other polynomials of degree 2 have a zero 1 or 0 and hence are divisible by $x - 1$ and x respectively.

In the general case, for arbitrary k, the prime polynomials have to be found by trial and error and there may be prime polynomials of arbitrarily high degree.

Note. Imitating Euclid's proof that there are infinitely many primes one easily proves that there are infinitely many prime polynomials in $k[x]$, k a field, but, as seen above, this statement does not say that there are prime polynomials of arbitrarily high degree.

R. Prove that $k[x]$ has prime polynomials of arbitrarily high degree when k has a finite number of elements.

A note on divisibility in integral domains

In the theory of divisibility in an integral domain I, one distinguishes between irreducible elements which do not permit factorization in nonunits and prime elements a with the property that if a divides a product, it divides at least one of the factors. The two need not coincide. The integral domain $Z + \sqrt{-5}Z$ has irreducible elements which are not primes since, for instance, $2.3 = (1 + \sqrt{-5})(1 - \sqrt{-5})$, where each factor turns out to be irreducible. But, provided that

(i) I does not have infinite sequences where each element is a proper (non-trivial) factor of the preceding one,

(ii) every irreducible element is prime,

the preceding proof goes through in I. In $k[x]$, (i) holds by the properties of the degree and (ii) holds as a consequence of the division algorithm. It is not difficult to show that the divisibility theorem holds in every commutative ring without zero divisors where (i) holds an every ideal is principal (*principal ideal ring*). A special case of these rings are the *Euclidean domains*, i.e. integral domains R with an abstract Euclidean algorithm, defined simply as the existence of a *degree function*, i.e., a function d from $R \setminus 0$ to the natural numbers with the property that $d(a) = 1$ when a is invertible, $d(a) < d(ab)$ for all a and b when b is not invertible, and that given non-zero elements a and b of R with $d(a) > d(b)$, there are elements q and r of R such that $a = qb + r$ and either $d(r) < d(b)$ or $r = 0$.

R. Prove in the same way as with $k[x]$ that if R is a Euclidean domain, then every ideal I of R is principal, $I = Ra$ for some a in I. Prove that if $Ra = Ra'$, then $a = ua'$ for some invertible element u. Finally, verify that R has the properties (i) and (ii) and hence has a unique factoriztion into primes.

Rings of the form

$$R = Z[\sqrt{d}]$$

whose elements are $a + b\sqrt{d}$ with a, b in Z, where d is an integer and not a square, may or may not possess a Euclidean algorithm and those not having one may not have a unique factorization into primes. As we know (see the section in Chapter 1 on the Gaussian integers), $d = -1$ gives the first case and it was indicated above that $d = -5$ gives the second. In algebraic number theory it is shown that unique factorization can be restored if one passes from primes and products of primes to ideals and products of ideals.

R. Prove that $Z[\sqrt{2}]$ is a Euclidean domain. (Hint. Try $d(a + b\sqrt{2}) = |a^2 - 2b^2|$.)

Prime polynomials in Q[x] *and* Z[x]

In Q[x] there are prime polynomials of arbitrary degree and no general irreducibility criteria. There is also the fact proved by Gauss that if a polynomial in Z[x] is prime (i.e., not the product of two polynomials also in Z[x] of positive degrees) then it is also prime in Q[x]. This is a consequence of the following two lemmas.

7.1 Divisibility in a polynomial ring

GAUSS'S LEMMA 1. *Let*

$$f(x) = a_0 + \cdots + a_n x^n$$

be a polynomial with integral coefficients. Suppose that a and b are integers with $(a,b) = 1$ and $f(a/b) = 0$. Then a divides a_0 and b divides a_n.

R. Prove the lemma by contemplating the following equation which follows from $f(a/b) = 0$,

$$a^n a_n + \cdots + b^n a_0 = 0.$$

With f as above, let $\text{cont} f$, the *content* of f, denote the greatest common divisor of its coefficients.

GAUSS'S LEMMA 2. *The content is a multiplicative function on polynomials with integral coefficients, i.e., one has*

$$\text{cont}(fg) = \text{cont} f \ \text{cont} g.$$

PROOF: It is clear that if a prime number divides all the coefficients of f or g, then it divides all the coefficients of fg. Hence it is enough to prove the lemma when the contents of f and g are 1. With f as above, let

$$g(x) = b_0 + b_1 x + \cdots + b_m x^m.$$

Suppose that p is a prime dividing the content of fg and let a_i and b_j be the first coefficients of f and g respectively that are not divisible by p. Then, if we compute mod p, (i.e., use the extension lemma with the map $a \to a$ mod p from Z to Z_p),

$$(fg)(x) \equiv a_i b_j x^{i+j} + \text{higher terms}.$$

Hence also the polynomial $f(x)g(x)$ mod p has a first coefficient not divisible by p. This is a contradiction and the lemma is proved.

R. Show that the second lemma implies the first one.

COROLLARY. *If f is a polynomial and prime in $Z[x]$, then it is also prime in $Q[x]$.*

PROOF: Let f be in $Z[x]$ and suppose that $f = gh$, with g and h in $Q[x]$. Write $g = ag'$ and $h = bh'$ where a and b are rational numbers and g' and h' have integral coefficients and content 1. Since $uf = vg'h'$ for some coprime integers u and v, Gauss's lemma 2 gives $u = v$ so that $f = g'h'$. Hence f being prime in $Z[x]$ implies that h' or g' equals 1 so that f is prime also in $Q[x]$.

The following criterion is well known.

EISENSTEIN'S CRITERION. *Suppose that*
$$f(x) = a_0 + \cdots + a_n x^n$$
is a polynomial with integral coefficients and that there is a prime p dividing all coefficients except a_n and that p^2 does not divide a_0. Then f is prime.

PROOF: Suppose that $f = gh$ where
$$g(x) = b_0 + \cdots + b_m x^m, \quad h(x) = c_0 + \cdots + c_q x^q,$$
where both degrees are positive. Since p but not p^2 divides $a_0 = b_0 c_0$, p must divide precisely one of the factors, say b_0. Let b_k be the first coefficient of g not divisible by p. There must be such a $k > 0$ since not all the coefficients of g are divisible by p. We have $k < m$ and, computing mod p, we get
$$f(x) \equiv b_k c_0 x^k + \text{higher terms}.$$
By assumption, p divides $b_k c_0$, so that p must divide c_0, which gives a contradiction.

Example
Let
$$f(x) = 1 + x + x^2 + \cdots + x^{p-1} = \frac{x^p - 1}{x - 1},$$
where p is prime. Then
$$f(x+1) = \frac{(x+1)^p - 1}{x} = x^{p-1} + p x^{p-2} + \cdots + p,$$
where, by the binomial theorem, the requirements of Eisenstein's criterion are fulfilled. Hence $f(x)$ is irreducible.

Note. The polynomials above belong to the class of *cyclotomic* polynomials to be treated in the next section.

Partial fractions

The decomposition of a rational function into partial fractions is probably well-known from analysis. Here we will treat the general case. The result to prove is

THEOREM. *Let $f(x)/g(x)$ be a rational function in $k(x)$, k a field. Write $g(x) = g_1(x)^{q_1} \ldots g_m(x)^{q_m}$ where $g_1(x), \ldots, g_m(x)$ are irreducible and different. Then there are uniquely determined polynomials $h(x)$ and $f_{ij}(x)$, $1 \leq j \leq q_i$, $1 \leq i \leq m$, such that $\deg f_{ij} < \deg g_i$ and*
$$\frac{f(x)}{g(x)} = h(x) + \sum_{i=1}^{m} \Big(\sum_{j=1}^{q_i} \frac{f_{ij}(x)}{g_i(x)^j} \Big).$$

R. Carry through the proof in two steps. First, using that the polynomials

$$G_i(x) = g(x)/g_i(x)^{q_i}$$

are pairwise prime, prove that polynomials $f_i(x)$ exist such that

$$\frac{f(x)}{g(x)} = h(x) + \sum_{i=1}^{m} \frac{f_i(x)}{g_i(x)^{q_i}}$$

and $\deg f_i < \deg g_i^{q_i}$ or $f_i = 0$. Then prove that the expansion is unique. Secondly, prove the following lemma using Euclid's algorithm.

LEMMA. *Let $f(x)$ and $g(x)$ be two polynomials in $k[x]$ and assume that $\deg g \geq 1$. Then there are uniquely determined polynomials $f_i(x)$, $i = 0, 1, \ldots, n$ such that $\deg f_i < \deg g$ and*

$$f(x) = f_0(x) + f_1(x)g(x) + \cdots + f_n(x)g(x)^n.$$

R. Let $f(x)$ and $g(x)$ be polynomials in $k[x]$, k a field, $n = \deg f > \deg g$. Prove that

$$\frac{g(x)}{f(x)} = \sum_{1}^{n} \frac{g(a_i)}{f'(a_i)(x - a_i)}$$

where $a_1, \ldots, a_n$ are the zeros of $f(x)$, assumed to be separate.

Exercises

1. Let k be a field and consider the set I of polynomials $f(x)$ in $k[x]$ such that $f(a) = 0$ for every a in k. Show that I is an ideal and that I is not 0 if and only if k is finite.
2. Show that $Z_2[x]$ has precisely two irreducible polynomials of degree 3 and that both are factors of $x^7 + 1$.
3. Factor $x^4 - x^3 + x^2 - x + 1$ into irreducible polynomials in $Z_2[x]$.
4. Factor $x^4 + x^3 + x^2 + x + 1$ into irreducible factors in $Z_3[x]$.
5. How many polynomials of the form $x^2 + ax + b$ are irreducible in $Z_2[x]$?

7.2 Algebraic numbers and algebraic fields

Let k be a field imbedded in an integral domain R, i.e., k is a subset of R, its algebraic operations are inherited from R and the 0 and 1 of R are also the 0 and 1 of k. Precisely as in the classical case when $k = Q$ and $R = C$, we say that an element a of R is *algebraic* over k if there is a polynomial $f(x)$ in $k[x]$ such that $f(a) = 0$, the computations being carried out in R. The proof of the first theorem of section 1.6 carries over to the new situation, which gives

THEOREM. *The algebraic elements over k form a field.*

R. Go through the proof in detail.

LEMMA. *Suppose k is a field embedded in an integral domain R and that a in R is algebraic over k. Then there is a unique irreducible monic polynomial $f(x)$ in $k[x]$ such that $f(a) = 0$.*

PROOF: Let $f(x)$ be a monic polynomial of lowest possible degree in $k[x]$ such that $f(a) = 0$. Then f must be irreducible, since otherwise there would be a proper factor g of f such that $g(a) = 0$. Suppose that $h(x)$ is another irreducible, monic polynomial in $k[x]$ such that $h(a) = 0$. Since $\deg f \leq \deg h$, we may divide h by f and get $h = qf + r$, where $\deg r < \deg f$ or r is the zero polynomial. But if r is not identically zero, then $r(a) = 0$, which contradicts the choice of f. Hence f divides h, which is possible only if $f = h$, since h was irreducible.

Existence

Let k be any field. We shall show how to construct in an explicit way *algebraic fields over k* (or *algebraic field extensions* of k), i.e., fields K in which k is embedded and whose elements are all algebraic over k. Note that if K is any field in which k is embedded, then K is a vector space over k. We denote the dimension of K as a vector space over k by $[K:k]$ and call it the *degree* of K over k. It can be finite or infinite. If finite, the extension is said to be *finite* also.

R. Show that if $[K:k] = n$ is finite, then K is an algebraic extension of k. (Hint. If x is an element of K, then $1, x, x^2, \ldots, x^n$ must be linearly dependent over k.)

Note. The converse of this is not true. The field of algebraic numbers discussed in Chapter 1 is an algebraic extension of the field of rational numbers, but of infinite degree.

LEMMA. *Suppose that K is an extension of a field k and that F is an extension of K and that both are finite. Then F is a finite extension of k and $[F:k] = [F:K][K:k]$.*

R. Prove the lemma. (Hint. If $u_1, \ldots, u_m$ is a basis for K over k and $v_1, \ldots, v_n$ a basis for F over K, show that $u_i v_j$ is a basis for F over k.

R. Prove that if either F is an infinite extension of K or K an infinite extension of k, then F is an infinite extension of k. Also prove the converse. Hence the lemma is valid also in this situation.

The following basic result shows how to construct algebraic fields.

THEOREM. *If $f(x)$ is an irreducible polynomial in $k[x]$ of positive degree and $(f(x))$ is the corresponding principal ideal $k[x]f(x)$, then the quotient*

$$K = k[x]/(f(x))$$

is an algebraic field over k. If u is the image of x under the quotient map, then $1, u, \ldots, u^{n-1}$ is a basis for K as a vector space over k, where $n = \deg f$. In particular, $[K : k] = n$.

PROOF: It is clear that K is a ring. Let $g(x)$ be any polynomial in $k[x]$ which is not divisible by $f(x)$. Since $f(x)$ is irreducible, this means that the greatest common divisor of $f(x)$ and $g(x)$ has degree < 1 and hence there are polynomials $a(x)$ and $b(x)$ such that

$$a(x)g(x) + b(x)f(x) = 1.$$

It follows that $a(x)g(x) \equiv 1 \mod f(x)$. Hence $a(x)$ is an inverse of $g(x)$ mod $f(x)$ so that K is a field. The restriction of the quotient map to k is clearly injective, so k is embedded in K. If $g(x)$ is a polynomial in $k[x]$, then we can write $g(x) = q(x)f(x) + r(x)$, where $\deg r < \deg f$ or else $r(x)$ is the zero polynomial. Hence the image of g in K equals the image of r. But this can clearly be written as a linear combination of $1, u, \ldots, u^{n-1}$ with coefficients in k. Finally suppose that $a_0 + a_1 u + \cdots + a_{n-1} u^{n-1} = 0$ for some a_i in k. Then $a_0 + a_1 x + \cdots + a_{n-1} x^{n-1} = g(x)f(x)$ for some polynomial $g(x)$. But since $\deg f = n$ this is impossible unless all a_i are zero.

Note. A simple-minded description of K runs as follows: the elements of K are polynomials in u with which one computes as usual with the rule $f(u) = 0$ added. Repeated applications of this rule allows one to write every element of K as a polynomial of degree $< n$.

Example

If k is the field of real numbers and $f(x) = x^2 + 1$, then K is the field of complex numbers and $1, i$ is a basis over R, where i is the image of x under the quotient map. This element is called the *imaginary unit*.

Splitting fields

The field constructed in the lemma above can be denoted simply by $k(u)$. In this field $f(x)$ has the zero u, so by the factor theorem it factors as

$$f(x) = (x - u)g(x),$$

where $g(x)$ has coefficients in $k(u)$. Write $g(x) = g_1(x) \ldots g_k(x)$, where the $g_i(x)$ are irreducible over the field $k(u)$. Now we can repeat our construction with one of the irreducible factors, say $g_1(x)$, getting a zero v of $g_1(x)$, and hence of $g(x)$, in a field $k(u)(v)$. Finally we arrive at a field F where $f(x)$ splits into linear factors,

$$f(x) = a(x - u)(x - v)(x - w)\cdots,$$

where a is the leading coefficient of f. In this way we have constructed a *splitting field* of $f(x)$, i.e., an extension of k over which $f(x)$ splits into linear factors and which has no proper subfields over which $f(x)$ splits into linear factors.

Note. In the definition of splitting field for a polynomial $f(x)$, this does not have to be irreducible. Clearly we can construct splitting fields for reducible polynomials also by using the method above. We only start by factoring $f(x)$ into irreducible factors and then treat these one by one.

The following theorem is important.

THEOREM. *Let k, k' be two fields and $\varphi : k \to k'$ an isomorphism. Let $f(x)$ be a polynomial in $k[x]$ and $\varphi f(x)$ the corresponding one in $k'[x]$. Suppose that F and F' are splitting fields for $f(x)$ and $\varphi f(x)$ over k and k' respectively. Then φ extends to an isomorphism of F and F'.*

The proof of this theorem hinges on the following

LEMMA. *Let k, k' be two fields and $\varphi : k \to k'$ an isomorphism. Suppose that $k \subseteq F$ and $k' \subseteq F'$ are extensions of k and k' respectively. Let $f(x)$ be an irreducible polynomial in $k[x]$ and $\varphi f(x)$ the corresponding one in $k'[x]$. Suppose that a is a zero of $f(x)$ in F and a' a zero of $\varphi f(x)$ in F'. Then the fields $k(a)$ and $k'(a')$ are isomorphic.*

R. Prove the lemma. (Hint. Verify that $\varphi(a) = a'$ gives the required extension of φ.)

PROOF OF THE THEOREM: We will prove the theorem by using induction over the degrees of F and F' over k and k' respectively. If $[F : k] = 1$, then $f(x)$ splits already over k, and there is nothing to prove. Assume the theorem is true for all splitting fields of degree less than $[F : k]$ and that $[F : k] > 1$. Then there is a factor $p(x)$ of $f(x)$ and a zero a of $p(x)$ in F which is outside k and a zero a' of $\varphi p(x)$ in F' also outside k. By the lemma φ extends to an isomorphism $\varphi : k(a) \to k(a')$. Now clearly F and F' are splitting fields for $f(x)$ over $k(a)$ and $k'(a')$ respectively. But $[F : k] = [F : k(a)][k(a) : k] > [F : k(a)]$ and by the induction hypothesis $\varphi : k(a) \to k'(a')$ extends to an isomorphism of F and F'.

7.2 Algebraic numbers and algebraic fields

Note. It can be shown that if $f(x) \in k[x]$ has degree n, then the degree of a splitting field over k is at most $n!$.

The following theorem has important applications.

THEOREM. *Let k be a field and let $f(x)$ and $g(x)$ be polynomials over k. Then $g(x)$ divides $f(x)$ if and only if $g(x)$ splits in a splitting field of $f(x)$ and every zero of $g(x)$ in this splitting field is also a zero of $f(x)$ and the multiplicity as a zero of $g(x)$ is at most the multiplicity as a zero of $f(x)$.*

PROOF: To prove "only if", we just write $f(x) = h(x)g(x)$. On the other hand, if $g(x)$ splits in a splitting field K of $f(x)$ and every zero of $g(x)$ in K is a zero of $f(x)$ of at most the same multiplicity, then clearly $f(x) = h(x)g(x)$, where $h(x)$ has coefficients in K. We must prove that they actually belong to k. Let

$$f(x) = a_0 + \cdots + a_n x^n, \quad g(x) = b_0 + \cdots + b_n x^n, \quad h(x) = c_0 + \cdots + c_n x^n$$

(where some coefficients presumably are zero). We get

$$b_0 c_0 = a_0$$
$$b_1 c_0 + b_0 c_1 = a_1$$
$$\cdots$$
$$b_n c_0 + \cdots + b_0 c_n = a_n$$

which is a quadratic system of linear equations with the c_i as unknowns and with coefficients in k. The corresponding homogeneous system has obviously only the trivial solution. Hence, by a theorem in the section on linear algebra, the system has a unique solution in k for all right sides. This proves the theorem.

Remark. If $f(x)$ is a polynomial over the field of complex numbers, then f can be factored as $f(x) = a(x-u)(x-v)\ldots$, where $u, v, \ldots$ are the zeros of $f(x)$ and a is the leading coefficient. Fields with this property are said to be *algebraically closed*. An example of a field which is not algebraically closed is the field R of real numbers, since the polynomial $x^2 + 1$ does not factor over R. If k is any field, it can be shown that there is an algebraically closed field $\bar{k}$, called the *algebraic closure* of k, such that every element of $\bar{k}$ is algebraic over k. The algebraic closure of the rationals Q is the field of algebraic numbers (which is *not* equal to the field of complex numbers).

Derivatives and multiple zeros

When k is an arbitrary field, one defines the derivative of a polynomial $f(x) = a_0 + a_1 x + \cdots + a_n x^n$ by the formula

$$f'(x) = a_1 + 2a_2 x + \cdots + n a_n x^{n-1}.$$

Then one has the basic properties

$$(f + g)'(x) = f'(x) + g'(x), \ (fg)'(x) = f'(x)g(x) + f(x)g'(x).$$

R. Prove these formulas by direct computation.

If $f(x)$ has degree n, i.e., $a_n \neq 0$, then $f'(x)$ has degree $n - 1$ provided that $na_n \neq 0$. This is always true in characteristic 0, but may not be true if the characteristic is $p > 0$. In spite of these complications we have

THEOREM. *An element u of a splitting field of $f(x)$ is a simple zero of f if and only if $f'(u) \neq 0$. The zeros of the greatest common divisor of $f(x)$ and its derivative $f'(x)$ are the zeros of $f(x)$ of multiplicity greater than one, provided that $f'(x)$ is not identically zero.*

PROOF: Write $f(x) = (x - u)^m g(x)$ where m is the multiplicity of u as a zero of $f(x)$ and $g(x)$ has coefficients in the splitting field and $g(u) \neq 0$. Since

$$f'(x) = (x - u)^{m-1}(mg(x) + (x - u)g'(x)),$$

the first assertion follows.

R. Prove the second assertion. Note that the greatest common divisor has coefficients in the field k.

COROLLARY. *When f is irreducible and f' is not identically zero, then all zeros of f are simple.*

Prime polynomials in characteristic p

When k has characteristic $p > 0$, it may happen that $f(x)$ has positive degree, but that $f'(x)$ is identically zero, for instance when $f(x) = x^p - 1$. This means that $ia_i = 0$ for all coefficients a_i of $f(x)$. This in turn is equivalent to the condition that $f(x) = g(x^p)$ where $g(x)$ is another polynomial with coefficients in k. When $g'(x)$ is identically zero, g in turn has this form and finally we arrive at a power p^t of p and a polynomial $h(x)$ whose derivative is not identically zero and such that

$$f(x) = h(x^{p^t}).$$

When f is irreducible, g is of course also irreducible. The point of all this is that irreducible polynomials may have multiple zeros if the field has prime characteristic. If, however, $k = Z_p$, we have $a^p = a$ for all a in k and, since $(u+v)^p = u^p + v^p$ in every field of characteristic p, the formula above reads

$$f(x) = g(x)^{p^t},$$

7.2 Algebraic numbers and algebraic fields

and it follows that irreducible polynomials have simple zeros. Since the previous theorem shows that this holds whenever $f'(x)$ is not identically zero, we have proved

LEMMA. *When a field k has characteristic zero or else $k = Z_p$, all irreducible polynomials in $k[x]$ have simple zeros.*

R. Let $f(x)$ be in $k[x]$ and u a zero of f in a splitting field. Using the formulas above, prove that u is a simple zero of f if and only if $f'(u) \neq 0$.

Application to cyclotomic polynomials

In this section we will work over the rational numbers, i.e., $k = Q$.
The complex zeros of the polynomial $f(x) = x^n - 1$ are the nth roots of unity,

$$\epsilon_j = e^{2\pi i j/n}, \ j = 0, 1, \ldots, n-1.$$

They form a cyclic group under multiplication whose generators, also called *primitive roots of unity*, are the ϵ_j with $(j, n) = 1$. There are $\varphi(n)$ of them.

R. Prove that if $f(x)$ and $g(x)$ have integral coefficients and $g(x)$ has leading coefficient 1, then there are polynomials $q(x), r(x)$ in $Z[x]$ such that $f = qg + r$ and $\deg r < \deg g$ or else $r(x)$ is the zero polynomial.

THEOREM. *There is an irreducible monic polynomial of degree $\varphi(n)$ with integral coefficients, the nth cyclotomic polynomial, whose zeros are the the primitive n^{th} roots of unity.*

PROOF: Define $f_n(x) = \prod_{(j,n)=1}(x - \epsilon_j)$. We want to show that f_n has integral coefficients and is irreducible.

First it follows that $x^n - 1 = \prod_{d|n} f_d(x)$. In fact, both sides are monic and have the same zeros. Also, $f_1(x) = x - 1$ and

$$f_n(x) = \frac{x^n - 1}{\prod_{d|n, d<n} f_d(x)}.$$

Hence the polynomials $f_d(x)$ can be computed recursively and since all $f_d(x)$ have leading coefficient 1, it follows from the R above that $f_n(x)$ has integral coefficients. It remains to prove irreducibility.

Let w be a primitive nth root of unity and $f(x)$ its minimal polynomial, i.e., the monic polynomial of lowest degree with rational coefficients such that $f(w) = 0$. This f is uniquely determined (why?) and is irreducible. We have to prove that $f = f_n$. In any case, f divides $x^n - 1$, so that $x^n - 1 = f(x)g(x)$, where by Gauss's lemma 2, both f and g have integral coefficients. We are going to show that every power w^p where p is a prime not dividing n, is a zero of $f(x)$. This suffices, because any primitive nth root of unity equals w raised to an exponent which is prime to n and every such exponent is a product of powers of primes which are prime to n.

The desired statement follows if we can prove that w^p cannnot be a zero of the factor $g(x)$ above. Suppose that $g(w^p) = 0$. Then w^p is a zero of $g(x^p)$ so that

$$g(x^p) = f(x)h(x)$$

for some monic polynomial $h(x)$ with integral coefficients. Now reduce this equality mod p. Then

$$g(x)^p \equiv f(x)h(x) \mod p.$$

Here the left side cannot be a factor of $h(x)$ mod p (why?). Hence $f(x)$ and $g(x)$ have a factor in common mod p. But then $x^n - 1 \equiv f(x)g(x)$ mod p has a multiple factor. Since $(x^n - 1)' = nx^{n-1}$ does not vanish mod p unless $x = 0$ mod p (remember that $(p,n) = 1$), this is a contradiction.

R. There is an explicit formula for the $f_n(x)$, namely

$$f_n(x) = \prod_{d|n}(x^{n/d} - 1)^{\mu(d)},$$

where $\mu(d)$ is the Möbius function. Prove this formula, which is essentially the Möbius inversion formula of section 1.3.

Exercises
1. Show that $x^2 - 2$ is irreducible over Q$[\sqrt{3}]$.
2. Determine an irreducible polynomial in Q$[x]$ with the zero $2 + \sqrt{i}$.
3. Determine an irreducible polynomial in Q$[x]$ with the zero $2 + \sqrt{-5}$.
4. Show that

$$Q + Q\sqrt{2} + Q\sqrt{3} + Q\sqrt{6}$$

is a field.

5. Let x and y be complex numbers such that $x^2 + x + 1 = 0$ and $y^2 + 3 = 0$. Which of the fields Q(x), Q(y) and Q$(\sqrt{3})$ are isomorphic?
6. Determine the field automorphisms of a) Q, b) Q$(\sqrt{i})$, c) Q$(\sqrt{2})$.
7. Prove that all extensions of the real numbers of degree 2 are isomorphic.
8. Let F be a finite extension of the real numbers of degree > 1. Show that $[F : R] = 2$ and that F is isomorphic to the complex numbers.

7.3 Finite fields

As we have seen before, Z_p is a field when p is a prime. In particular Z_p is a *finite* field, i.e., it has a finite number of elements. It is not difficult to construct *all* finite fields (up to isomorphism).

If F is a finite field, it must have prime characteristic p in which case all integral multiples of the unit 1 form a field F_0, the *prime field* of F and isomorphic to Z_p. If the degree $[F : F_0]$ of F over F_0 is n, then F must have $q = p^n$ elements.

THEOREM. *If u is an element of the finite field F with q elements, then $u^q = u$. In particular, if $u \neq 0$, then $u^{q-1} = 1$.*

PROOF: Let F^* be the set of non-zero elements of F. If u is in F^*, then the map $x \to ux$ is a bijection of F^*. Hence $u^{q-1}Q = Q$, where Q denotes the product of all the elements of F^*. Since $Q \neq 0$, this proves the theorem.

Note. If $F = Z_p$, this can be expressed as $a^{p-1} \equiv 1 \mod p$ for all non-zero elements of F, which is Fermat's little theorem.

Existence

It is easy to construct the field F above. In fact, let F be the splitting field of the polynomial
$$f(x) = x^{p^n} - x$$
over Z_p. Since its derivative $f'(x) = p^n x^{p^n - 1} - 1 = -1$ identically, all the zeros are different. We shall see that they form a field.

R. Using the identity
$$(a+b)^p = a^p + b^p,$$
valid in any ring of characteristic p, show that if u and v are zeros of $x^{p^n} - x$, then so are $u - v$ and uv. Show also that the inverse of a non-vanishing zero is a zero.

Hence the splitting field must consist precisely of the zeros of $x^{p^n} - x$, so it has p^n elements. On the other hand, if F' is another field with p^n elements, then $x^{p^n} - x$ splits over F' also, so F and F' are isomorphic. We have proved

THEOREM. *A finite field has prime characteristic p and p^n elements for some natural number n. It is isomorphic to the splitting field of $x^{p^n} - x$.*

Note. The field we have constructed is called a *Galois field* of order p^n after its discoverer. It will be denoted by $GF(p^n)$ in the sequel.

Primitive elements and primitive prime polynomials

THEOREM. *In every Galois field F there is a primitive element, i.e., an element u whose powers generate all of F except the zero.*

PROOF: The non-zero elements of F form an abelian group. From Lemma 1 of section 3.4 we know that the order of any element divides the maximal order m. Hence every non-zero element of F is a zero of the polynomial $x^m - 1$. Since this polynomial has at most m zeros in any splitting field, m must be the number of elements of $F \setminus 0$ and this proves the theorem.

Note. In group theoretic terminology: the multiplicative group of any Galois field is cyclic.

R. Let k be a field, not necessarily finite, and M a finite subgroup of $k \setminus 0$. Show that M is cyclic.

R. Let u be a primitive element of $GF(p^n)$ and $f(x)$ its minimal polynomial in $Z_p[x]$. Show that its degree equals n. Hence $GF(p^n)$ is isomorphic to $Z_p[x]/(f(x))$. In other words, there are irreducible polynomials of every degree over Z_p.

Example
Let $f(x) = x^2 + 1$ be in $Z_3[x]$. Since $f(a)$ is not zero for any a in Z_3, f is a prime polynomial. Hence, if u is a zero of f in a splitting field, $Z_3(u)$ is a Galois field with 9 elements.

Note. A polynomial $f(x)$ in $Z_p[x]$ which has a zero that is a primitive element for $GF(p^n)$ will be called a *primitive prime polynomial*.

Automorphisms of finite fields

THEOREM. *A finite field F with p^n elements has precisely n automorphisms. They are given by $a \to a^{p^m}$ for $m = 0, 1, \ldots, n-1$.*

PROOF: Since the prime field Z_p consists of the multiples $0, e, 2e, \ldots, (p-1)e$ of the unit e, any automorphism must leave the prime field elementwise fixed. Since $(a+b)^p = a^p + b^p$, $(ab)^p = a^p b^p$ for all a, b in F, the map $T : u \to u^p$ is an endomorphism. It is not the zero map, so its kernel is zero (for F is a field). Since F is finite, T must be an automorphism. Then all powers T^k are also automorphisms.

Conversely, let S be an automorphism of F and let c be a primitive element of F. Then $S(c)$ is also a primitive element and hence $S(c) = c^r$ for some integer $r > 0$ and $< p^n$ and prime to $p^n - 1$. It follows that the same equation holds for all powers of c and hence $S(a) = a^r$ for all a in F. Write $r = r'p^k$ where p does not divide r'. Then $U = ST^{-k}$ is another automorphism with the property that $U(a) = a^{r'}$ for all a. The equality

$U(1 + a) = 1 + U(a)$ means that the polynomial $(1 + x)^{r'} - 1 - x^{r'}$ of degree $r' - 1 < p^n - 1$ has p^n zeros, namely all the elements of F. Hence it is identically zero, which means in particular that $r'a = 0$ for all a when $r' > 1$, i.e., that $r' \equiv 0 \mod p$. This is a contradiction and the theorem is proved.

Exercises

1. Show that
$$(a + b)^{p-1} = \sum_{k=0}^{p-1}(-1)^k a^k b^{p-1-k}$$
in every field of characteristic p.

2. Let K be a field of characteristic $p > 0$. Find all polynomials $f(x)$ in $K[x]$ such that $f(x + y) = f(x) + f(y)$ for all x, y in K.

3. Write down a multiplication table for a field with four elements.

4. Show that the sum of all elements of a finite field with more that two elements vanishes.

5. Show that the map $x \to x^2$ of a finite field is bijective if and only if its characteristic is two.

6. Show that $GF(3^4)$ does not have a subfield with 27 elements but one with 9 elements.

7. Show that $x^m - 1$ divides $x^n - 1$ if and only if m divides n.

8. Show that $GF(p^n)$ has a subfield with p^m elements if and only if m divides n. (Hint. Use exercise 7.)

Literature

The material of this chapter can be found in most textbooks on algebra. Galois discovered the finite fields. His paper on the subject is from 1830.

CHAPTER 8

Shift registers and coding

Two of the most striking applications of finite fields to computer science are to the theory of shift registers and to coding. Both are of great practical importance. Here we will be concerned only with elementary theory.

8.1 The theory of shift registers

Periodical sequences

Functions a from the integers ≥ 0 to some set will be called right sequences, $a = (a(0), a(1), \ldots)$, and functions from all integers will be called full sequences, $b = (\ldots b(-1), b(0), b(1), \ldots)$. A *period* of a full sequence is defined to be any integer $m \neq 0$ such that $b(j + m) = b(j)$ for all j, a period of a right sequence is defined by the same equalities with the restriction that $m > 0$ and $j \geq 0$. Sequences possessing periods are said to be periodic.

R. Let $m > 0$ be a period of a right sequence a. Prove that m is also a period of a unique associated full sequence b defined by $b(j) = a(j)$ when $j \geq 0$ and $b(-km + j) = a(j)$ when $k > 0$ and $j = 0, \ldots, m - 1$.

LEMMA. *The periods of a right sequence are all positive integral multiples of the least period.*

PROOF: By the previous exercise, every period of a right sequence is also a period of the associated full sequence. For positive periods, the converse is trivial. Now it is obvious that 0 and all periods of the full sequence form a module over the integers and hence consists of all integral multiples of a positive integer m which is also the least positive period of a.

R. The sequence $a = (1, 2, 3, 4, 1, 2, 1, 2 \ldots)$ is not periodic but it is ultimately periodic in the sense that it has *ultimate* periods, i.e numbers $m \neq 0$ such that $a(j + m) = a(j)$ for all sufficiently large j. Prove that 0 and all ultimate periods of a sequence form a module and hence that they consist of all integral multiples of a least positive period.

8.1 Shift registers

Shift registers

A *shift register* is a device (for instance a computer program) which computes the elements of a right sequence $a = (a(0), a(1), \ldots)$ with elements in some set S in such a way that the *input* $a(0), \ldots, a(n-1)$ is given in advance and the rest of the sequence is computed by the rule that

$$(1) \qquad a(j+n) = f(a(j+n-1), \ldots, a(j)).$$

for all $j \geq 0$. Here f is a fixed function called the *generating* or *feedback* function or simply the *feedback* of the shift register. A device which computes the function f is turned into a shift register if it is completed by a shift of the input one step to the right. This is the origin of the term shift register. Such devices have found many applications in all kinds of automata.

R. Assume that the set S is finite. Prove that any sequence generated by a shift register is ultimately periodic. (Hint. Suppose that the input has n elements. The sequence a is then the union of parts $A(0), A(1), \ldots$ where $A(k)$ consists of all $a(j)$ with $kn \leq j < (k+1)n$ and it is obvious that the register induces a map T such that $A(1) = TA(0)$ and, in general, $T^k A(j) = A(j+k)$ for $j, k \geq 0$. Prove that there are numbers $j < k$ such that $A(j) = A(k)$ and deduce from this that a is ultimately periodic with period $k - j$.

Algebraic theory of linear feedbacks

A feedback function is said to be linear when it generates sequences with elements in a field F and is itself linear and homogeneous,

$$(2) \qquad a(n) = c(1)a(n-1) + c(2)a(n-2) + \cdots + c(n)a(0)$$

with coefficients in F. The sequences generated by a linear feedback have a very precise algebraic theory. In this theory, a right sequence $a = (a(j))$ is made to correspond to a formal power series

$$(3) \qquad a(t) = \sum_{0}^{\infty} a(k) t^k$$

in one indeterminate t and a feedback given by (2) is made to correspond to the polynomial

$$(4) \qquad f(t) = 1 - c(1)t - \cdots - c(n)t^n,$$

the *feedback polynomial*. Any polynomial with constant term 1 can serve in this capacity. Under the correspondences (3) and (4), every sequence generated by the feedback f corresponds to the formal power series

$$a(t) = b(t)/f(t)$$

where $b(t)$ is a polynomial of degree $< n$. In fact, by (2), the coefficients of t^j in $a(t)f(t)$ vanish when $j \geq m$ and the rest is just a polynomial of degree $< m$. We can now prove

THEOREM. *A formal power series $a(t)$ has the period m if and only if*

(5) $$a(t) = b(t)/(1-t^m)$$

for some polynomial $b(t)$ of degree $< m$. When $f(t)$ is a polynomial and $f(0)=1$, the formal power series $1/f(t)$ has the period m if and only if $f(t)$ divides $1 - t^m$.

Note. The last condition implies in particular that all the zeros of $f(t)$ in a splitting field are roots of unity. When the field is finite, this is no restriction. In fact, then the splitting field is a Galois field all of whose elements except 0 are roots of unity.

PROOF: If a power series $a(t)$ is periodic with period m, then

$$a(t) = b(t) + t^m b(t) + t^{2m} b(t) + \cdots$$

where $b(t)$ is the sum of the first m terms of $a(t)$. Hence $a(t)$ has the form announced. The converse is obvious. This proves the first part of the theorem. To prove the second part, note that $1/f(t) = b(t)/(1-t^m)$ means that $f(t)b(t) = 1 - t^m$.

R. Prove that a formal power series $a(t)$ with coefficients in a field has an ultimate period if and only if $(1 - t^m)a(t)$ is polynomial $b(t)$, i.e. $a(t) = b(t)/(1 - t^m)$ with no restriction on the degree of $b(t)$.

R. Let $f(t)$ be a feedback polynomial of degree n and let $b(t)$ be a polynomial of degree $< n$. Prove that $1/f(t)$ and $b(t)/f(t)$ have the same least period if and only if $f(t)$ and $b(t)$ are coprime.

Least periods of feedback polynomials

When $f(t)$ is a feedback polynomial, let $\mathrm{per} f$ denote the least period of the formal power series $1/f(t)$.

8.1 Shift registers

THEOREM. *The least period of a polynomial $f(t)$ with $f(0) = 1$ and coefficients in Z_p, p a prime, is $p^j(p^k - 1)$ where p^j is the least power of p exceeding the largest multiplicity of a zero of $f(t)$ and $GF(p^k)$ is its splitting field. The maximal least positive period among polynomials of a given degree d is $p^d - 1$ and occurs for primitive irreducible polynomials.*

PROOF: According to the previous theorem, the least period is also the least number $m > 0$ such that $f(t)$ divides $1 - t^m$. Hence $m + 1$ is at least equal to the degree p^k of the splitting field of $f(t)$. Since all multiplicities of a polynomial of the form $1 - t^m$ only occur when m is a multiple of p and $(1 - t^{mp}) = (1 - t^m)^p$, the first statement of the theorem follows. To prove the second, note that if the degree of a polynomial is constant and its largest multiplicity decreases by one, then the degree of its splitting field must increase. Hence, since $p^j(p^k - 1)$ can only increase when j decreases by one and k increases, a polynomial of a given degree with maximal least period cannot have multiple zeros. It follows that the degree of its splitting field must be maximal. This proves the theorem.

Long periods and random numbers

When $f(t)$ is a primitive prime polynomial of degree n with coefficients in Z_p, the previous theorem tells us that the corresponding formal power series $1/f(t)$ has the period $q - 1$ where $q = p^n$. It turns out that the corresponding sequence, say

$$A = (a(0), a(1), a(2), \ldots),$$

considered as a circular one, has remarkable homogeneity properties which makes it a good candidate for producing random numbers. For simplicity we put $p = 2, q = 2^n$. Let T be the translation $a(k) \to a(k + 1)$ in A.

THEOREM. *With the assumptions above, let B be any $k \leq n$ elements within a set of n consecutive elements of A. Then a sequence of k consecutive zeros appears $2^{n-k} - 1$ times and the others 2^{n-k} times in the sequence $B, TB, \ldots, T^{q-1}B$.*

Note. The theorem indicates that if we let n successive elements of the sequence denote n-digit numbers, then we are reasonably sure that successive such numbers are reasonably independent.

Example

$f(t) = t^3 + t + 1$ is a primitive prime polynomial with coefficients in Z_2. The corresponding sequence A with initial elements 1,1,1 is the sequence $1, 1, 1, 0, 0, 0, 1, \ldots$ and the successive translates of the initial sequence are

$$1,1,1 \quad 0,0,1 \quad 1,0,0 \quad 0,0,1 \quad 0,0,1,0 \quad 1,0,1 \quad 0,1,1$$

i.e. all combinations of three zeros and ones except 0,0,0. In any place there are 4 ones and 3 zeros. The polynomial $t^{35} + t^2 + 1$, which is primitive and irreducible, has been used in practice to generate pseudorandom numbers on machines whose numerical register holds 35 digits. The corresponding period is infinitely long for practical purposes.

PROOF OF THE THEOREM: When an initial sequence $a(0), \ldots, a(n-1)$ is transported around the sequence A, its consecutive translates run through all sets of n binary numbers except the set consisting only of zeros. It follows that the values taken by by the sequence of the theorem are those taken by k elements in fixed places when n binary numbers are combined in all possible ways with the exception of the set consisting only of zeros. Counting these possible ways completes the proof of the theorem.

8.2 Generalities about coding

Generally speaking, a code C is a bijection T from a set A of signs, called letters to another set B of letters. A message M written with the letters of A is encoded by T to a coded message TM written in the letters of B and the original message is recovered by applying T^{-1} to TM, $M = T^{-1}TM$. There are of course many codes. In a classical example A and B consist of the same letters and T is a permutation of them.

Encoding messages is done with different aims. One of them is secrecy, the encoding should be difficult to encode. The other is reliability against errors of encoding, transmission and decoding. This is the origin of the important error-correcting codes. The utility of shift registers in both cases stem from their ability to encode and decode suitably constructed codes with great speed and without the aid of a large memory.

Block codes

A widely used class of codes are the block codes where the encoded message consists of code words w of fixed length (number of letters) forming a subset of all words of the same length. Each code word has a message part m which can be chosen arbitrarily and a check part c which is used to complete the message part to a proper code word. If an error of transmission has occurred, resulting in a word u which is not a code word, it is then possible to see that that there has been an error and it may be possible to locate the original code word as the closest code word to u.

For bit codes, i.e. codes where the letters are 0 and 1, we can define a distance $d(w, w')$ between two words w and w' as the number of digit positions where the two words differ.

R. Prove that this distance, called the Hamming distance, is symmetric,

8.2 Generalities about coding

$d(w, w') = d(w', w)$, and satisfies the triangle inequality $d(w, w) \le d(w, u) + d(u, w')$.

The minimal distance between two different code words is called the *separation* of the code. The distance of a word to a code is defined to be the least distance to a code word.

LEMMA. *If the separation of a code is odd, say $2d+1$, then any word whose distance to the code is d has a unique nearest neighbor in the code.*

R. Prove this lemma using the triangle inequality.

We shall consider codes, the block codes, in which the words, the message and the check have, respectively, a fixed number of binary digits, say $n, k, j = n - k$. They are referred to as (n, k)-codes. When the separation d is exhibited, the notation (n, k, d) is used. For economic codes the separation should not be too small compared to the length of the check.

Example
The classical *parity check* code has just one check digit and the requirement that the sum of the digits of a code word should be even. This code does not correct errors but it discovers that some error has been committed.

Linear codes

The parity check is an example of a *linear* code where the digits of of the code words belong to some finite field F, for instance Z_2 in case of a binary code, and constitute a linear subspace in a linear space of finite dimension over a finite field. This subspace is identified with the code. The basic properties of linear codes, stated in the exercises below, follow from the basic facts of abstract linear algebra (see section 5.3).

R. Let C be a linear subspace of dimension k in F^n, F a field. Prove that, after a suitable permutation of the digit positions, C has a basis consisting of vectors $e(1), \ldots, e(k)$ whose first k components are zero except the jth component of $e(j)$ which is one. In this way, the matrix with rows $e(1), \ldots, e(k)$ has the form (I, A) where I is the unit $k \times k$ matrix and A is some $k \times (n - k)$ matrix.

R. Let C be a linear n-bit code. Prove that the Hamming distance d is invariant under translation, i.e. that $d(u + w, v + w) = d(u, v)$ for all code words u, v, w. It follows from this that the separation of the code is also the minimal distance between zero and non-zero code words. Prove that the separation of the linear (5,3) binary code defined by

$$a(4) = a(1) + a(3), \quad a(5) = a(1) + a(2) + a(3)$$

is 2.

R. Define the *weight* of a code word in a binary code to be the number of ones that it contains. Using the matrix (I, A) of basis elements of a linear (n, k, d) bit code, prove the Singleton bound: $d \leq n - k + 1$.(Hint. The basis elements have weight at most $n - k + 1$.)

Note. The Singleton bound is just one of many restrictions for error-correcting codes of type (n, k, d). Others are reviewed in the last section of this chapter.

R. Let C^* the dual space of C, be all vectors $t = (t(1), \ldots, t(n))$ which are orthogonal to C in the sense that the *interior product*

$$t(1)a(1) + \ldots t(n)a(n)$$

vanishes for all words a in C. Prove that C^* is a linear space of dimension $n - k$ and that C consists of all vectors orthogonal to C^*. The code C^* is said to be dual to C.

8.3 Cyclic codes

A famous class of linear codes are the *cyclic* n-bit codes with the property that if the code contains a word

$$w = (a(0), \ldots, a(n-1))$$

it also contains the cyclically shifted word Sw defined by

$$w \to Sw = (a(n-1), a(0), \ldots, a(n-2)).$$

R. Prove that the dual code of a cyclic code is cyclic. (Hint. If (u, v) is the interior product introduced above, prove that $(Su, v) = (u, S^{-1}v)$.)

Cyclic codes turn out to have nice properties when expressed in polynomial form. For this, let every word $w = (w(0), \ldots, w(n-1))$ of given length n correspond to a polynomial

$$w(t) = w(0) + w(1)t + \cdots + w(n-1)t^{n-1}$$

in the quotient ring $T_n = GF(2)[t]/((t^n - 1))$, obtained by adding and multiplying polynomials under the rule that $t^n = 1$. In particular, we can restrict ourselves to polynomials of degree $< n$. Using our polynomial notation, we have $(Sw)(t) \equiv tw(t)$. Iterations of this formula and the linearity prove that the code is cyclic if and only if the corresponding polynomials constitute an *ideal* I in T_n. If $g(t)$ is a non-zero polynomial of lowest degree in I (and hence unique since its coefficients are binary digits), the division algorithm for polynomials shows that I consists of all polynomial multiples of $g(t)$. Then $t^n - 1 = h(t)g(t) + r(t)$ where $\deg r(t) < \deg g(t)$ and this is only posssible when $r(t) = 0$, i.e. $g(t)$ divides $t^n - 1$. Hence we have proved

8.3 Cyclic codes

LEMMA. *Every cyclic (n,k) code in polynomial form consists of all multiples mod $t^n - 1$ of a unique polynomial $g(t)$ of degree $n - k$ dividing $t^n - 1$.*

Note. The polynomial $g(t)$ is said to *generate* the code.

Example

The Hamming (7,4) code has the generating polynomial $1 + t + t^3$ which is irreducible in $Z_2[t]$. Its zeros are u, u^2, u^4 where u generates $GF(8)$. This code has the separation 3 and hence corrects one error. To see this, suppose that two code word polynomials differ in at most two places. Then their difference $w(t)$ is a polynomial with at most two terms so that $w(t) = t^m$ or $w(t) = t^m + t^n$ where $0 \leq m < n < 7$. Since $w(u) = 0$, the first case is impossible and, with $k = n - m$, the second case gives $1 + u^k = 0$ which is impossssible since the order of u is 7.

The previous lemma reduces some basic properties of cyclic codes to simple algebraic exercises.

R. Let C_1 and C_2 be binary cyclic codes of length n with generating polynomials $g_1(t)$ and $g_2(t)$. Prove that C_1 contains C_2 if and only if $g_2(t)$ divides $g_1(t)$.

R. Prove that a binary cyclic code C with generating polynomial $g(t)$ has no word of even weight if and only if $t + 1$ divides $g(t)$. (Hint. Prove that a word has even length if and only if the corresponding polynomial is divisible by $t + 1$.)

R. Prove that a binary cyclic code possessing at least one word of odd length contains the word whose letters are all ones. (Hint. Prove that there is a polynomial $f(t)$ such that $1 - t^n = (t - 1)f(t)g(t)$ and that $f(t)g(t)$ corresponds to the desired word.)

R. Prove that the interior product of two words u and v of the same length is the constant term of the product $u(1/t)v(t)$.

R. Let $g(t)$ generate a cyclic (n, k) code C and let $h(t)$ be the cofactor of $g(t)$ in $1 - t^n$. Then $h(t) = h(0) + h(1)t + \cdots + h(k)t^k$ has degree k. Let

$$h^*(t) = t^k h(1/t) = h(k) + h(k-1)t + \cdots + h(0)t^k$$

be the polynomial *reciprocal* to $h(t)$. Prove that $h^*(t)$ generates the dual code of C. (Hint. Prove that all constant terms of the products

$$t^{-s}h^*(1/t)t^r g(t)$$

vanish when $0 \leq s < n - k$ and $0 \leq t < k$.)

Encoding and decoding of cyclic codes

The encoding of a cyclic (n, k) code can use an n-bit shift register. In fact, the generating polynomial $g(t)$ gives a sequence

$$g = (g(0), \ldots, g(n-k-1), 0, \ldots, 0)$$

and the polynomial $t^j g(t)$ corresponds to the sequence $S^j g$ which is the sequence g shifted j steps to the right. If the message is represented by the polynomial $m(t) = \sum m(j) t^j$ of degree $\leq k$, the encoded sequence is $\sum m(j) S^j g$ corresponding to the polynomial $w(t) = g(t) m(t)$. Let $h(t)$ be the cofactor of $g(t)$ in $1 - t^n$. Decoding is then performed by computing the product $h(t) w(t)$ modulo $1 - t^n$. In fact, $h(t) w(t) \equiv m(t)$. Checking is done by verifying that w is orthogonal to all shifts of the word corresponding to the reciprocal polynomial of $h(t)$.

Idempotent generators and quadratic residue codes

Let C be a cyclic n-bit code with generator polynomial $g(t)$ and define $h(t)$ by $g(t)h(t) = t^n - 1$. If n is odd, the derivative of the right side vanishes only when $t = 0$ and hence its zeros are separate. It follows that $g(t)$ and $h(t)$ are coprime and hence there are polynomials $a(t)$ and $b(t)$ such that $1 = a(t)g(t) + b(t)h(t)$. Here $e(t) = a(t)g(t)$ is in the ideal generated by $g(t)$ and multiplication by $g(t)$ shows that $g(t) \equiv a(t)g^2(t)$ so that $g(t)$ is also in the ideal generated by $e(t)$. Further, $e(t) \equiv e^2(t)$ so that $e(t)$ is also *idempotent*. We have proved that cyclic n-bit codes have idempotent generators when n is odd.

Idempotent generators are useful in the construction of *quadratic residue* codes. They are cyclic but they are also invariant under certain additional shifts. They are p-bit codes where p is a prime. If

$$a = (a(0), \ldots, a(p-1))$$

is in the code, and t is a quadratic residue mod p, it is required that also the sequence $(b(0), \ldots, b(n-1))$ where $b(j) = a(k)$ with $k \equiv tj$ mod p is also in the code. We shall present the two simplest codes of this kind.

Let p be a prime and let Q and N be the quadratic residues mod p and the quadratic non-residues respectively. Assume further that $p^2 \equiv 1$ mod 8 which means, in particular, that 2 is a quadratic residue mod p. The polynomials

$$Q(t) = \sum t^j, \quad N(t) = \sum t^j$$

for j in Q and N respectively, are then idempotents modulo $t^p - 1$ for the map $j \to 2j$ is a permutation both of Q and of N. Let (Q) and (N) be the corresponding codes generated by $Q(t)$ and $N(t)$. We shall see that they have interesting properties.

THEOREM. *When $p \equiv -1(8)$, (Q) and (N) have the dimension $(p-1)/2$ and a common separation d for which $d^2 > p$.*

PROOF: By hypothesis, $(p-1)/2$ is odd and hence, by an earlier exercise of this section, the unit polynomial $H(t) = \sum_0^{p-1} t^k$ is an idempotent generating a code (H) belonging to both (Q) and (N). This means that $H(t) \equiv C(t)Q(t)$ for some $C(t)$ so that $H(t)Q(t) \equiv Q(t)$ and the analogous formula for $N(t)$. Combining this with $H(t) = 1 + Q(t) + N(t)$ shows that $H(t) \equiv Q(t)N(t)$, i.e. $(H) = (Q) \cap (N)$. Note that $t^j H(t) \equiv H(t)$ for all j so that $f(t)H(t) \equiv H(t)$ for all polynomials $f(t)$ which means that $H(t)$ is the only element in (H).

Next consider the additive automorphism of Z_p induced by a change of sign, $j \to -j$. It induces a multiplicative bijection $S : t^j \to t^{-j}$ of the powers of t modulo $t^p - 1$, which extends to a ring morphism $T^p \to T^p$ where T^p is the ring $Z_p[t]$ modulo the ideal generated by $t^p - 1$. Since $p \equiv -1 \bmod 4$, -1 is not a quadratic residue and this means that $j \to -j$ sends quadratic residues into non-residues and conversely. Hence S sends Q into N and vice versa. Now $\dim((Q) \cap (N)) = 1$ and $\dim(Q) = \dim(N) = p$ for any polynomial $f(t)$ satisfies

$$f(t) \equiv f(t)(H(t) + Q(t) + N(t))$$

where, inside the parenthesis, $H(t) \equiv Q(t)N(t)$, so that $(Q) + (N)$ is all of T^p. Since (Q) and (N) overlap in just one dimension, $p = 2\dim(Q) - 1$ so that $\dim(Q) = \dim(N) = (p-1)/2$.

To prove the last part of the theorem, note that the separations of (Q) and (N) are the same. This follows since S permutes the two. Let d be their common separation and let $f(t)$ and $g(t)$ be members of (Q) and (N) of weight d. Then $f(t)g(t)$ belongs to $(Q) \cap (N)$ and hence is $\equiv H(t)$ so that the weight of $f(t)g(t)$ is at least p. Since it is obvious that this same weight is at most d^2, we have proved that $d^2 \geq p$. Since equality cannot hold, this finishes the proof of the theorem.

8.4 The BCH codes and the Reed-Solomon codes

The proof given above that the Hamming (7,4) code has separation 3 extends to certain other cyclic codes, the BCH codes, after their inventors Bose, Chaudhauri and Hoquenghem. The following lemma is used.

LEMMA. *Let F be a finite field and let u be a primitive nth root of unity. If a polynomial of degree $\leq n$ with coefficients in F vanishes for d successive powers of u, it has at least $d + 1$ non-vanishing coefficients.*

PROOF: Let $f(t) = \sum a(k) t^k$ with k running over d separate integers $r(0), \ldots, r(d-1)$. Suppose that it vanishes for d successive powers of

$u, u^s, u^{s+1}, \ldots, u^{s+d-1}$. Put $v(p) = u^{r(p)}$ and $b(p) = a(r(p))v(p)^s$ with $p = 0, \ldots, d-1$. Then the equations $f(u^{s+q}) = 0$ for $q = 1, \ldots, d-1$ give us a system of equations as follows

$$\sum_p b(p)v(p)^q = 0$$

for $p, q = 0, \ldots d-1$. Since u is a primitive nth root of unity, the algebraic numbers $v(0), \ldots, v(d-1)$ are all different so that, by the properties of the van der Monde determinant, (see the next exercise), all coefficients $b(p)$ must vanish. This proves the lemma.

R. Prove van der Monde's formula

$$\det(v(j)^k)_{j,k=0,\ldots,n-1} = \prod_{j<k}(v(j) - v(k)).$$

(Hint. Subtract suitable multiples of the first row (where $j = 0$) from the others to get zeros in the first column except at the top. Take out the factors $v(k) - v(0)$ for $k = 1, \ldots, n-1$ and contemplate the result.)

Application

Consider binary codes and let u be a primitive element of $GF(2^n)$. Also, let $I(u,t)$ denote the unique irreducible polynomial which has a zero u. Then the polynomial

$$f(t) = \mathrm{LCM}(I(u,t), I(u^2,t), \ldots, I(u^{2k},t))$$

where $2k < 2^n$ vanishes for all powers of u between u and u^{2k} and hence serves as the generating polynomial for a cyclic code with separation $2k+1$, which hence corrects k errors. The check length is the degree of $f(t)$. The word length must of course exceed that number.

R. Prove that the degree of $f(t)$ is at most $2k$. (Hint. If $f(v) = 0$ then $f(v^2) = 0$ so the problem amounts to proving that at most k polynomials are needed in the product. Use induction.)

Examples

$k = 1, u = 1$ gives $f(t) = 1$ so that we get the parity check code. The case $k = 2, n = 3$ is realized by $f(t) = t^3 + t + 1$ which gives the Hamming (7,4) code. In fact, $f(t)$ is irreducible and if u is a zero so is u^2 and they are separate. More examples are available in Macwilliams-Sloane (1977).

Reed-Solomon codes

So far most of our code words have been written with binary letters. They can be replaced by the elements of a finite field, for instance n-tuples of zeros and ones representing the elements of the field $F = GF(2^n)$. With this remark in mind, let u be a generator of the cyclic group $F\backslash 0$ and let $f(t)$ be the polynomial

$$(t - u)(t - u^2)\ldots(t - u^{2j})$$

where $2j < 2^n$. Considered as a polynomial with coefficients in F, it has $2j$ zeros which are consecutive powers of one element. Hence, by the arguments above used in the construction of BCH codes, $f(t)$ is the generating polynomial of a code with letters in F which corrects j errors. If we write the letters as binary n-tuples, and consider binary errors, we have a code which corrects a burst of consecutive binary errors as long as the burst does not cover more that j letters of F. Hence the maximal length of a correctable burst is $n(j-1) + 1$.

Example
When $n = 8, j = 5$, the code corrects 5 errors when written in terms of F. Hence the code corrects bursts of 33 binary errors.

8.5 Restrictions for error-correcting codes

When C is a (n, k, d) code let $R = k/n$ be its *rate* and $\delta = d/n$ be its relative separation. The Singleton bound (see the end of section 8.2), says that $R + \delta \leq 1 - 1/n$. There are some other formulas of the same kind. Two of them use a function

$$F(n, k, q) = \sum_{i=0}^{k} \binom{n}{i}(q-1)^i.$$

1) *Sphere-packing bound.* Let C be a code of length n using q letters and define the Hamming distance beween two words as the number of places where they differ. Then a sphere of radius r around a given word w contains

$$V(r) = F(n, r, q)$$

words (and not only code words). In fact, the ith term of the sum $F(n, r, q)$ is the precise number of words which differ from w in i places. Now, if the code corrects t errors, then the spheres of radius t around all q^k code words cover at most all q^n possible words and this gives the Hamming inequality (1950) or sphere-packing bound

$$V(t)q^k \leq q^n.$$

Taking the q-logarithm of both sides gives the following estimate of the rate when t is given

$$R \leq 1 - n^{-1} \log V(t).$$

This is a numerical version of the fact that the rate decreases when t increases and the word length is fixed.

2) *Varshamov-Gilbert bound*

R. Let C be a linear (n,k) block code with letters in a finite field F, let C^* be the dual code consisting of all n-vectors $t = (t(1), \ldots, t(n))$ with coefficients in F wich are orthogonal to C in the sense that

$$\sum_{k=1}^{n} w(k)t(k) = 0$$

for all n-vectors $w = (w(1), \ldots, w(n))$ corresponding to words in C. By the theory of abstract linear algebra, C^* has a basis B which we can think of as a $(n-k) \times n$ matrix M, called the check matrix of C. Prove that C consists of all n-vectors orthogonal to the elements of B and that C has the minimal weight d if and only if any $d-1$ columns of M are linearly independent.

R. Prove that there are $F(n,i,q) - 1$ non-zero linear combinations (with coefficients in a field with q elements) of at most i out of n vectors.

The announced estimate, due to Gilbert and Varshamov, says that as long as

$$F(n-1, d-2, q) < q^{n-k},$$

it is possible to construct a linear (n,k) code with coefficients in $GF(q)$ and minimal weight at least d. The proof relies on the preceding exercises. By the first exercise, it suffices to construct appropriate check matrices. Suppose that we have already constructed a $(n-1, k-1, d-1)$ code with the desired properties in form of a check matrix M of the type $(n-k) \times (n-1)$ where any $d-1$ columns are linearly are independent. We shall try to add another column to M. We then have to avoid $F(n-1, d-2, q) - 1$ non-zero linear combinations of the already constructed columns and add another one if we can find one. Now, since $n \geq k$, the total number of non-zero columns available in the extended check matrix is $q^{n-k} - 1$. Hence the desired result follows by induction.

8.5 Restrictions

Stated in another way, our bound says that

$$R \geq 1 - n^{-1} \log F(n-1, d-2, q).$$

3) *Plotkin bound.* For binary (n, k, d) codes where $d < n/2$, the rate is extremely limited. This follows from the Plotkin bound

$$2^k \leq 2d/(2d-n).$$

To prove this, consider the sum S of the distances between all pairs of non-identical code words. Since all these distances are at least d, S is at least $dK(K-1)$ where $K = 2^k$. If we write the code words as as vectors $(u(1), \ldots, u(n))$, and arrange them into a matrix M, S equals

$$\sum_{u \neq v} \sum_{i} d(u(i), v(i)).$$

Summing over i fixed gives the result $L = 2x(i)(k - x(i))$ where $x(i)$ is the number of zeros of M in the ith column. Since L is at most $K^2/2$, and since there are n columns we get

$$2dK(K-1) \leq K^{2n}$$

which is the desired result.

Asymptotic bounds

The origin of coding theory was Shannon's theorem about transmission in a noisy binary channel. He proved that if the probability of transmission error per one bit signal is p, then there are (n, k) codes with rates as close to $1 - H(p)$ as one wants which, when n is sufficiently large, permit a transmission with arbitrarily small error probability. Here $H(p)$ is the entropy function

$$H(p) = p \log \frac{1}{p} + (1-p) \log \frac{1}{1-p},$$

log being $\log_2$. The entropy vanishes when $p = 0, 1$ and has a maximum 1 when $p = 1/2$. In the theory of error-correcting codes, there is no complete analogue of this result, but the entropy function appears in the asymptotic forms of the bounds above when $q = 2$. Estimating the binomial coefficients by Stirling's formula, one arrives at the following asymptotic bounds for R as a function of $\delta = d/n$,

$R = 0$ when $\delta \geq 1/2$, (Plotkin)

$R \leq 1 - H(\delta/2)$, (sphere packing).

The asymptotic Varshamov-Gilbert bound asserts the existence of arbitrarily long codes for which

$R \geq 1 - H(\delta)$ when $\delta \leq 1/2$.

There are further necessary conditions which limit the pair (R, δ) to a crescent-shaped region sketched in the figure below. A good family of codes is currently defined as one containing arbitrarily long codes for which the rate exceeds the Varshamov-Gilbert bound. One such family is the Goppa codes, invented by Goppa in 1970 and further developed by the use of algebraic geometry over a finite field (see Lachud 1985).

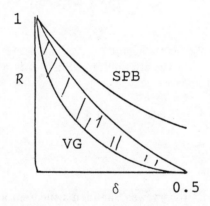

Figure. The asymptotic region for good codes (shaded) lies above the asymptotic Varshamov-Gilbert bound (VG) and below certain other bounds including the sphere-packing bould (SPB).

Literature

The first part of the chapter dealing with shift registers is based on a series of lectures (in Swedish) given at Lund University by T. Herlestam in 1979. Most of the theory of algebraic coding appeared first in the technical literature and has had a rapid development from 1950 on. The Theory of Error-correcting Codes by MacWilliams and Sloane (1977) is the present bible of the subject. It is very clear and well written with about 1500 references. A more elementary book dealing also with certain combinatory aspects not covered here is Vera Pless 1982.

CHAPTER 9

Groups

One of the uses of group theory is to classify symmetries. It has therefore been argued, although with more enthusiasm than logic, that group theory is as old as the symmetric decorative patterns appearing in various cultures, for instance islam.

The concept of a group as we know it was known to Lagrange, Abel and Galois, but it did not gain general importance until Camille Jordan wrote a book (1870) explaining the until then obscure writings of Galois. Groups became important in physics with the arrival of quantum physics and the necessity to study the symmetries of atomic orbitals. Nowadays, group theory is part of a general education in mathematics. This chapter ends with a short section on the applications of group theory to combinatorics.

9.1 GENERAL THEORY

The object of the first part of this chapter is to review the essentials of the general theory of groups. It has many features and part of the terminology in common with the theory of modules in the form of abelian groups.

9.1.1 Groups and subgroups

A *group* is a non-empty set $G = \{a, b, c, \ldots\}$ with multiplication and division. In other words, there is a function $(a, b) \to ab$ from $G \times G$ to G called the product (or composition) such that

(i) $(ab)c = a(bc)$ for all a, b, c in G (associativity),

(ii) G has a unit (identity element) e with the property that $ae = ea = a$ for all a in G,

(iii) every a in G has an inverse, i.e. an element b in G such that $ba = ab = e$.

Note. The inverse of a is written a^{-1}.

If the multiplication is commutative, $ab = ba$ for all a, b in G, the group is said to be *commutative* or *abelian*. Apart from the multiplicative notation, abelian groups are the same as the Z-modules of Chapter 2.

R. Show that there is only one unit element and that no element of a group has two different inverses.

Examples

We have already seen many groups, e.g. all modules. The set of non-zero elements of a field or a division ring form multiplicative groups and also the positive elements of Q and R.

R. Show that the complex numbers of absolute value 1 form a commutative group under multiplication.

R. Show that the set of all bijective affine functions in one variable, i.e. real functions f of the form $f(x) = ax + b$, $a \neq 0$, is a group under composition of functions (the affine group in one variable). Write down the inverse of f and the unit element. Is the group abelian?

Show that this group is also represented by pairs (a, b) of real numbers, $a \neq 0$, with the law of composition

$$(a, b) \cdot (c, d) = (ac, ad + b).$$

A *subgroup* of a group G is a non-empty subset H of G which is itself a group under the multiplication in G.

R. Show that a non-empty subset H of a group G is a subgroup if and only if

$$a, b \in H \Rightarrow ab^{-1} \in H.$$

R. Show that the intersection of two subgroups of a group is again a subgroup.

R. Show that the affine functions with i) $a > 0$, ii) $a = 1$ (the translations), iii) $b = 0$ (the dilations) are subgroups of the affine group in one variable. Use this to show that the union of two subgroups need not be a subgroup.

Any group G has some natural subgroups, e.g. its *center* cent(G) consisting of all elements a of G which commute with all elements b, $ab = ba$. For every a in G there is also its *centralizer* $C(a)$ consisting of all elements b which commute with a. While the center is always abelian, there is no reason why centralizers should be.

R. Verify that the center and all centralizers are subgroups and that the center is the intersection of all centralizers. Show that the center of the affine group above consists of the unit element alone.

R. Let G be the set of all invertible $n \times n$ matrices with entries in a field. Show that G is a group (under matrix multiplication). Show that the center of G consists of all matrices of the form λI, where λ is a scalar and I is the unit matrix.(Hint. Use triangular matrices.)

9.1.1 Groups and subgroups

Generators

R. Let $ab\ldots c$ be a product of elements of a group G. Show that its inverse is $c^{-1}\ldots b^{-1}a^{-1}$.

This exercise shows that every non-empty subset M of a group G generates a subgroup, namely the set of all products $a_1 a_2 \ldots a_n$ where a_i or a_i^{-1} is in M.

Cyclic groups

Any element a of a group G generates what is called a *cyclic* subgroup, namely the set
$$\{a^n; n \in \mathbb{Z}\}$$
of all powers, positive or negative or zero, of a, the power with exponent zero being the unit element.

R. Show that a cyclic group is either infinite in which case all the powers a^n are different, or that there is an integer $m > 0$ such that the group has m elements, namely all powers of a with exponent ≥ 0 and $< m$.

R. *Roots of unity.* The nth roots of unity, i.e. the complex numbers z whose n^{th} power is 1, form a natural abelian group of order n (i.e. it has n elements). Show that it is cyclic and that the kth power of a generator is a generator if and only if k and n are coprime. Generators of this group are also called *primitive roots of unity* (see also Chapter 7). Show that their number is $\varphi(n)$, where φ is Euler's function.

Direct product

Corresponding to the direct sum of modules, we have the direct product $G \times H$ of two groups. Its elements are pairs (a,b), where $a \in G$ and $b \in H$. The product is defined by
$$(a_1, b_1)(a_2, b_2) = (a_1 a_2, b_1 b_2).$$

R. Prove that the direct product of two groups is a group. Write out the unit and inverses.

Exercises

1. Show that all functions from the complex plane of the form $f(z) = az + b$ where $a^n = 1$ and b is an arbitrary complex number form a group. Determine its center. Determine the centralizer of the element $g(z) = z + b$, $b \neq 0$.

2. Let G be a group generated by four elements a, b, c, d for which $ab = c, bc = d, cd = a, da = b$. Show that G is cyclic of order 5.

9.1.2 Groups of bijections and normal subgroups

Let $X = \{x_1, x_2, \ldots\}$ be a set. All bijections $f : X \to X$ from X to itself constitute a group under composition:

$$(fg)(x) = f(g(x)).$$

The identity is the function $x \to x$ for all x and the inverse of f is the function $f(x) \to x$.

If X has a finite number of elements $x_1, \ldots, x_n$ a bijection is just a permutation of X: the elements

$$f(x_1), \ldots, f(x_n)$$

are the elements $x_1, \ldots, x_n$ in a certain order.

Note. The group of all bijections of a set is also called the permutation group of the set regardless of how many elements the set has. Subgroups of such groups are universal examples of groups. In fact, we shall see later that every group is essentially such a subgroup.

R. Show that the following conditions define subgroups in *every* group of bijections $X \to X$ (and not only the group of all bijections):
1) all f with $f(x) = x$ for a fixed x in X (the *stabilizer* of x),
2) all f with $f(Y) = Y$ where Y is a fixed subset of X.

Examples

In the affine group in two variables, i.e. the group of affine functions in two real variables x and y, those of the form

$$f(x, y) = (ax + by, cx + dy),$$

with real a, b, c, d and $ac - bd \neq 0$, the functions that map the circle

$$x^2 + y^2 = 1$$

onto itself constitute a group, the group of rotations about the origin and reflections in lines through the origin. Analytically, such functions can be written as

$$f(x, y) = (\epsilon(\cos\theta x - \sin\theta y), \sin\theta x + \cos\theta y),$$

where $\epsilon = 1$ means a rotation and $\epsilon = -1$ a reflection.

Exercise

Verify explicitly that the elements of the affine group in one variable whose parameters a, b satisfy an equation $at + b = 0$ with t a fixed real number form a subgroup.

Normal subgroups

A subgroup H of a group G is said to be *normal* if

$$a \in H, b \in G \Rightarrow bab^{-1} \in H$$

or, what is the same thing, $bHb^{-1} = H$ or $bH = Hb$ for all b in G. All subgroups of an abelian group are obviously normal, but in the general case normal subgroups are rather special.

R. Show that the group of translations is a normal subgroup of the affine group in one variable but that the subgroup of dilations is not normal. Generalize to the affine group in two variables.

R. Let G be that group of invertible $n \times n$ matrices with entries in a field. Show that the subset of all matrices with determinant 1 is a normal subgroup, but that the subgroup of matrices with zeros in the first row and the first column except that the top left element is 1 (or not zero) is not normal when $n > 1$.

Groups of *linear bijections* (linear transformations) occur in linear algebra.

R. Let M be a vector space of dimension n over a field k. Show that all bijective linear maps from M to itself form a group. This group is called the general linear group of dimension n over k, $\mathrm{GL}(n, k)$. The special cases $k = \mathrm{R}$ and $k = \mathrm{C}$ are among the so-called classical groups. In general one thinks of the elements of these groups as invertible $n \times n$ matrices with entries in k.

Exercises

1. Let H and K be normal subgroups of a group G and assume that H and K have only the unit in common. Show that the elements of H and K commute. (Hint. Rewrite $hk = kh$ as $h^{-1}k^{-1}hk = e$.)

2. Which are the finite subgroups of the multiplicative group of the non-zero complex numbers?

3. Suppose that the equation $x^2 = e$ in a group has precisely one solution $x \neq e$. Show that x belongs to the center of the group.

4. Let R be a non-commutative ring with a unit e and let U be the group of invertible elements of R. Let I be a two-sided ideal of R and V the set of elements x of U such that $e - x$ is in I. Show that V is a normal subgroup of U.

9.1.3 Groups acting on sets

As we have seen, there are plenty of groups consisting of bijections of a set X onto itself. Such groups are also very frequent and occur in situations which can be formalized as follows.

A group $G = \{a, b, \ldots\}$ is said to *act* on a set $X = \{x, y, \ldots\}$ if there is a mapping $G \times X \to X$ written as $(a, x) \to a.x$, such that $e.x = x$ for all x in X, e being the unit of G, and $a.(b.x) = (ab).x$ for all a, b in G and x in X.

Note. We have written the 'product' $a.x$ with a point to distinguish it from the product ab of G. When the context shows what product is meant, we shall sometimes leave out the point.

R. Show that these conditions imply that the map $x \to a.x$ from X to itself is a bijection f_a for all a in G, that $f_{ab} = f_a f_b$ in the sense of composition and that f_e is the identity map. (Hint. The value of f_a at x is $f_a(x) = a.x$.)

When a group G acts on a set X, every element x of X gives rise to a subgroup of G, namely its *stabilizer*, stab(x), consisting of all elements a in G such that $a.x = x$.

R. Show that the intersection of all stabilizers is a normal subgroup. It follows that the center of a group is a normal subgroup.

R. Let H be a subgroup of a group G and define a product $a.b$ with a in H and b in G by
(i) $a.b = ab$, (ii) $a.b = ba^{-1}$, (iii) $a.b = aba^{-1}$.
Show that all of them define actions of H on G.

Orbits

Let the group G act on the set X. An *orbit* of G in X is a subset of X defined by
$$G.x = \{g.x; g \in G\}$$
with a fixed element x of X. This set is also written as orb(x) and is called the orbit through x.

R. Show that if x is in orb(y), then orb(x) = orb(y).

R. Let G be the group of rotations of a plane about a point. Describe the orbits of G in the plane. Describe the orbits of the group generated by one plane translation. Let G be the group generated by the function $f(x) = -x + b$, $b \neq 0$ and real. Show that every orbit of G acting on the real line has precisely two elements and that two orbits which have a point in common are identical.

Our last exercise illustrates the following simple theorem which is basic in group theory.

9.1.3 Groups acting on sets

THEOREM. *When a group G acts on a set, the orbits form a partition of it, i.e. the set is the disjoint union of the orbits.*

PROOF: Every x in the set X lies in at least one orbit, namely $\text{orb}(x)$. If an element z of X lies in the orbits $\text{orb}(x)$ and $\text{orb}(y)$, then by the exercise above, $\text{orb}(x) = \text{orb}(y)$, since both are equal to $\text{orb}(z)$.

R. A group G acting on a set X is said to act *transitively* when X is the only orbit. Prove that the affine group acts transitively on the real line.

Conjugacy classes

Let G be a group and define an action of G on itself by

$$a.x = axa^{-1}.$$

Its orbits are called *conjugacy classes* and two elements in the same orbit are said to be *conjugate*. In particular, G is the disjoint union of conjugacy classes. Conjugacy classes are only interesting for non-abelian groups, because every element of an abelian group is itself a conjugacy class.

R. Show that a conjugacy class has only one element if and only if this element belongs to the center of the group. Show that the stabilizer of an element under the action above is its centralizer.

Cosets

A subgroup H of a group G acts on the bigger group by the formula

$$a.x = ax$$

with a in H and x in G. The orbits of this action are important subsets of G of the form

$$Hx = \{ax; a \in H\},$$

called *right cosets* of H in G. *Left cosets* of H are the subsets xH of G, orbits under the left action of H on G.

R. If $H = \{e\}$, every element of G is a coset. Show that if H and K are subgroups of G and K is contained in H, then the orbits under H are partitioned by the orbits under K.

The theorem above has the following important consequence.

THEOREM. *Any group is the disjoint union of the left (or the right) cosets of any subgroup.*

A subset M of G is called a *set of representatives* for a given subgroup when it contains precisely one element from every coset of the subgroup, left or right as the case may be.

Group morphisms and quotient groups

The theorems on module morphisms and quotient modules in the theory of modules have counterparts for groups. First we state the definitions.

A mapping $f : G \to H$ from one group to another is said to be a *group morphism* or a *homomorphism* if

$$f(ab) = f(a)f(b)$$

for all a, b is G. (Note that the multiplication on the right is that of the group H.) An *isomorphism* is a bijective homomorphism.

Example

The map $x \to e^x$ is an isomorphism from R with addition to the positive reals with multiplication.

R. Show that two cyclic groups with the same number of elements, finite or infinite, are isomorphic. It follows that a cyclic group of finite order n is isomorphic to Z/nZ and that one of infinite order is isomorphic to Z with addition.

R. Let $f : G \to H$ be a group morphism. Show that the *image* of f, $\text{im} f = f(G)$ is a subgroup of H. Define the *kernel* of f, $\ker f$, as the set of elements of G which are mapped by f into the unit of H. Show that $\ker f$ is a *normal* subgroup of G.

R. When $f(x) = ax + b$ is an affine bijection and $g(x) = ax$, show that $f \to g$ is a homomorphism whose kernel is the group of translations of R.

R. Show that a group morphism $f : G \to H$ is injective if and only if $\ker f$ consists only of the unit of H.

Quotient groups

If H is a normal subgroup of a group G, then by definition, $aH = Ha$ for all a in G. In other words, a subgroup is normal if and only if every left coset is also a right coset.

R. Verify this, i.e. show that $aH = Hb \Rightarrow aH = Ha$.

When M and N are subsets of G define their product MN as the set of elements ab of G with a in M and b in N. When H is a normal subgroup this gives

$$HaHb = H^2 ab = Hab$$

(why is $H^2 = H$?). Hence the product of two cosets is another coset. It is clear that $HaH = Ha$ and $HaHa^{-1} = H$.

R. Show that this multiplication of cosets is associative.

9.1.3 Groups acting on sets

All this shows that, if H is a normal sugroup of a group G, its cosets form a group under multiplication. This group is called the *quotient group* of G modulo H and is denoted by G/H precisely as for modules and rings.

Note. One can say that 'subgroup' corresponds to 'subring' in ring theory and that 'normal subgroup' corresponds to 'ideal'.

R. Show that the map $a \to aH$ from G to G/H is a group morphism which is an isomorphism if and only if H reduces to the unit element.

R. What is the quotient group when G is the group of affine functions in two variables and H the subgroup of translations?

R. Show that, for the same group, the subgroup of elements leaving the origin fixed is not a normal subgroup.

The following theorem is analogous to the module morphism theorem and it is proved in the same way.

GROUP MORPHISM THEOREM. *Let $f : G \to H$ be a group morphism. Then $G/\ker f$ is isomorphic to $\operatorname{im} f$.*

PROOF: Define a new function $\bar{f} : G/\ker f \to \operatorname{im} f$ by putting $\bar{f}(x\ker f) = f(x)$. Then $\bar{f}$ is well defined, for if $x\ker f = y\ker f$, then $y^{-1}x$ is in $\ker f$, so that $f(y^{-1}x)$ is the unit of H and hence $f(x) = f(y)$. It is injective, for if $\bar{f}(x\ker f) = 1$, then x is in $\ker f$ and so $x\ker f = \ker f$. Finally, the map $\bar{f}$ is clearly surjective.

R. Show that G is isomorphic to $\operatorname{im} f$ if f is injective.

The following two terminologically complicated exercises are optional.

R. Let H and K be two normal subgroups of a group G and suppose that K is contained in H. Prove that $(G/K)/(H/K)$ is isomorphic to G/H. (Hint. Consider the map $xK \to xH$ from G/K to G/H and use the morphism theorem.) The result of this exercise is called the second group morphism theorem.

R. Let H be a normal subgroup of a group G and K another subgroup, not necessarily normal. Show that the product HK is a group and that HK/K is isomorphic to $H/H \cap K$ (the third group morphism theorem).

Exercises

1. Show that $x \to e^{2\pi i x}$ is a group morphism from the real numbers R to the set T of complex numbers of absolute value 1 and that T is isomorphic to R/Z.

2. Let S be a non-trivial subgroup of a group G. Show that the complement of S generates G.

3. Let G be a group where every element a satisfies an equation $a^n = e$ for some integer n. Prove that every homomorphism f from G to the complex numbers has the property that $|f(a)| = 1$ for all a.

4. Let $\hat{C}$ be the complex plane completed by a point ∞ at infinity. Prove that the functions
$$f(z) = \frac{az+b}{cz+d},$$
where a, b, c, d are complex numbers such that $ad - bc = 1$, form a subgroup of all bijections of $\hat{C}$ (the Moebius group). Write down the inverse of f.

5. The *commutator subgroup* G' of a group G is the subgroup generated by all elements of the form $ghg^{-1}h^{-1}$. Show that G' is a normal subgroup and that the quotient group G/G' is abelian. Also show that if H is another normal subgroup of G such that G/H is abelian, then $G' \subseteq H$.

9.2 FINITE GROUPS

In this section all groups are assumed to be finite, i.e. to have a finite number of elements.

9.2.1 Counting elements

The number of elements of a set X is called its *order* and will be denoted by $|X|$. By the order of an element a of a group G we mean the order of the cyclic subgroup generated by a. This number n is also called the *period*, since all elements of the sequence of powers of a appear again after n steps. The number of cosets of a subgroup H of a group G is called the *index* of H in G and denoted by $[G : H]$.

R. Prove that a subgroup H whose index is 2 is normal. (Hint. G is the disjoint union of the cosets of H.)

The basic fact about finite groups is

THEOREM. *All cosets of a subgroup H of a finite group G have the same number of elements as H.*

PROOF: If two members ab and ac of a coset aH are equal, b and c must be equal.

Since the cosets form a partition of the big group we have the following classical result which shows that the theory of finite groups and number theory are connected.

COROLLARY (LAGRANGE 1775). *The order of a subgroup divides the order of the group. Their quotient is the index of the subgroup.*

9.2.2 Symmetry groups and the dihedral groups

R. Let a be an element of order n of a group G. Show that n divides $|G|$. Hence $a^{|G|} = e$.

Note. When G is the group of invertible elements of the ring Z/nZ, this gives the Euler-Fermat theorem.

R. Let p be a prime. Determine all groups of order p.

The following important theorem is a variant of the corollary.

THEOREM. *Suppose that a finite group G acts on a set X. Then*

$$|G| = |\text{stab}(x)||\text{orb}(x)|$$

for all x in X.

PROOF: It suffices to show that the number of cosets of $\text{stab}(x)$ is the same as the order of $\text{orb}(x)$. Consider the map

$$a\text{stab}(x) \to a.x.$$

of cosets of $\text{stab}(x)$ into $\text{orb}(x)$. This is really a map, since if $a\text{stab}(x) = b\text{stab}(x)$, then $b^{-1}a$ is in $\text{stab}(x)$, so that $b^{-1}a.x = x$ and $a.x = b.x$. Since $a.x = x$ if and only if a is in $\text{stab}(x)$, the map is injective. Surjectivity is evident and the theorem is proved.

The class formula

When x is an element of a group G, let $\text{Cl}(x)$ be the conjugacy class of x, i.e. the orbit of x under the action

$$y \to aya^{-1}$$

(conjugation) of the group on itself. The stabilizer of x under this action is the centralizer $C(x)$ of x. By the theorem above,

$$|\text{Cl}(x)| = \frac{|G|}{|C(x)|}.$$

If we combine this with the observation that every element of the center $\text{cent}(G)$ of G is its own conjugacy class, we arrive at the *class formula*, which is very important in the analysis of the structure of finite groups.

THE CLASS FORMULA. *For every finite group,*

$$|G| = |\text{cent}(G)| + \sum \frac{|G|}{|C(x)|}$$

where the sum runs over a set of representatives x of the conjugacy classes with more than one element.

PROOF: The formula merely expresses the fact that G is the disjoint union of the conjugacy classes.

9.2.2 Symmetry groups and the dihedral groups

Euclidean geometry, treated in Euclid's Elements (500 B.C.) and taught to generations of schoolchildren for at least four centuries, is the study of how objects in Euclidean space behave under congruence transformations. Analytically, Euclidean n-space is real n-space where the distance between two points x and y is defined by

$$|x-y| = (\sum_{k=1}^{n}(x_k - y_k)^2)^{\frac{1}{2}}.$$

A *congruence transformation* is an affine function $x \to f(x)$ in n variables leaving all distances invariant, $|x - y| = |f(x) - f(y)|$ for all x and y. Necessary and sufficient for this is that the homogeneous part of f be given by an orthogonal matrix.

The *symmetry group* of an object A in Euclidean space is the subgroup of all congruence transformations leaving A invariant in the sense that $f(A) = A$. Many symmetry groups have beeen studied, for instance those of the platonic regular solids, i.e. the tetrahedron, the cube (hexahedron), the octahedron, the dodecahedron, and the icosahedron with, respectively, four, six, eight, twelve, and twenty sides. In physics, crystal lattices are studied via their symmetry groups.

Here we shall limit ourselves to the study of the symmetry group of the plane n-gon, called the *dihedral group* D_n because the n-gon is considered to be situated in Euclidean 3-space and hence to have two sides. For simplicity, our treatment relies on ordinary geometrical intuition.

Label the vertices $0, 1, \ldots, n-1$ so that all vertices $i, i+1$ are connected by a side. Let a be a rotation by $2\pi/n$ around the center of the n-gon. Then $a^n = e$, where e is the identity, and $a(k) = k + 1$ if $0 \le k < n - 1$ and $a(n - 1) = 0$. The orbit orb(0) of 0 under the action of the cyclic group generated by a consists of all the vertices.

Let b be a reflection in the symmetry axis through 0, or, which is the same thing, a half-way turn around this axis in 3-space. Then $b^2 = e$ and

$$a^k b a^{-k}(j) = a^k b(j - k) = a^k(n - j + k) = n - j + 2k,$$

so that $a^k b a^{-k}$ is a reflection in the symmetry axis through k. Hence a and b generate D_n. Since $bab(j) = ba(n - j) = b(n - j + 1) = j - 1 = a^{-1}(j)$, we have $bab = a^{-1}$.

9.2.3 The symmetric and alternating groups

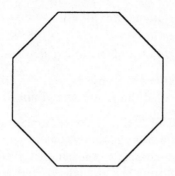

A regular octagon

R. Show that the relation $bab = a^{-1}$ implies that every element of D_n can be written $a^k b^m$, where $0 \leq k < n$, $0 \leq m < 2$. Hence D_n has order $2n$.

Note. The 2-gon should be considered to have two sides. If it is considered to be just an interval, its symmetry group, isomorphic to the cyclic group C_2 of order 2, is generated by a rotation half turn or, which gives the same result, a reflection in the midpoint.

Exercises

1. Let H and K be subgroups of a group G of orders m and n respectively and assume that $H \cap K$ has order k. Show that the set HK has mn/k elements.

2. Show that there is an even number of elements of a group equal to their inverses when the group has even order. (Hint. Group the elements in pairs (x, x^{-1}).)

3. Let G be a finite group. Show that the following conditions are equivalent:
(i) $|G|$ is odd
(ii) every element of G has odd order
(iii) the equation $x^2 = a$ has a solution for all a in G
(iv) the equation $x^2 = a$ has a *unique* solution for all a in G
(v) the equation $x^2 = e$ has the unique solution $x = e$.
(Hint. Prove (i) $\Rightarrow$ (ii) $\Rightarrow$ (iii) $\Rightarrow$ (iv) $\Rightarrow$ (v) $\Rightarrow$ (i).)

4. Let G be a finite group in which the equation $x^n = e$ has at most n solutions for every n. Show that G is cyclic. (Hint. Let C be a cyclic group of order $|G|$. Let $f(k)$ and $g(k)$ be the number of elements of G and C respectively of order k. Show that $f(k) = g(k)$ by first showing that

$f(k) \leq g(k)$ and then summing over all k.)

9.2.3 The symmetric and alternating groups

The group of permutations of n objects, e.g. the first n natural numbers, is an important group called the *symmetric group* S_n.

R. Prove that $|S_n| = n!$. (Hint. Use induction and the fact that one integer can move to n places by permutations.)

R. Show that S_3 is isomorphic to the symmetry group of an equilateral triangle. (Note that rotating the triangle around a symmetry axis is an operation in 3-space.)

R. Why is the symmetry group of a square a *proper* subgroup of S_4?

R. Prove *Cayley's theorem*: every group of order n is isomorphic to a subgroup of S_n. (Hint. Show that $a \to f_a$, f_a defined by $f_a(x) = ax$ with a and x in G, is an injective homomorphism from G to the group that permutates the elements of G.)

To compute the product (i.e. the composition) of two permutations of $1, \ldots, n$, one writes down their graphs,

$$\begin{pmatrix} 1 & \ldots & n \\ f(1) & \ldots & f(n) \end{pmatrix}, \begin{pmatrix} 1 & \ldots & n \\ g(1) & \ldots & g(n) \end{pmatrix}$$

and operates as follows: k is mapped to $g(k)$ by g which is mapped to $f(g(k))$ by f.

Example

$$f = \begin{pmatrix} 1 & 2 & 3 & 4 \\ 2 & 3 & 1 & 4 \end{pmatrix}, g = \begin{pmatrix} 1 & 2 & 3 & 4 \\ 4 & 3 & 2 & 1 \end{pmatrix}$$

Here $1 \to 4 \to 4$, $2 \to 3 \to 1$, $3 \to 2 \to 3$, $4 \to 1 \to 2$. The graph of the product is

$$fg = \begin{pmatrix} 1 & 2 & 3 & 4 \\ 4 & 1 & 3 & 2 \end{pmatrix}.$$

Cycles

To get a convenient, compact notation for a permutation f of $1, \ldots, n$, one writes down the set of orbits of the cyclic group generated by f. These orbits are called *cycles*. For example, if

$$f = \begin{pmatrix} 1 & 2 & 3 & 4 & 5 & 6 & 7 \\ 3 & 6 & 4 & 1 & 7 & 2 & 5 \end{pmatrix}$$

9.2.3 The symmetric and alternating groups

the cycle from 1, $(1, f(1), f^2(1), \ldots)$ is (134), that from 2 is (26) and that from 7 is (57) and we get

$$f = (134)(26)(57)$$

which means that $f(1) = 3, f(3) = 4, f(4) = 1, f(2) = 6, f(6) = 2, f(5) = 7, f(7) = 5$. Cycles with only one element is sometimes not written out in this notation. Cycles with two elements are called *transpositions*. Note that if $g = (134), h = (26), l = (57)$, then $f = ghl$, where the factors commute since they permute disjoint sets of integers.

R. Compute fg when f is as above and $g = (124)$.

It is clear from the example above that every permutation is the product of the cycles it generates and that the cycles commute.

R. Show that the permutation

$$(g(k), gf(k), gf^2(k), \ldots) \ldots$$

is equal to gfg^{-1} when g is another permutation. Show that every cycle and hence every permutation is a product of transpositions.

The preceding exercise (see the last formula) shows that the conjugacy class of a permutation f is characterized by the orders or lengths of its cycles. These can be written down in the form of a *cycle index*

$$(c_1, \ldots, c_n)$$

where c_i denotes the number of cycles of length i of f. Note that the numbers c_i are restricted by the condition that $\sum i c_i = n$.

Example

The cycle indices of the permutations $f = (124)(3567)$ of S_8 and $g = (12)(34)(567)$ of S_7 are, respectively,

$$(0, 0, 1, 1, 0, 0, 0, 0) \text{ and } (0, 2, 1, 0, 0, 0, 0).$$

The cycle index of the identity of S_n is $(n, 0, \ldots, 0)$.

It is not difficult to write down an explicit formula for the number of elements of a conjugacy class with a given cycle index.

LEMMA. *The number of elements of the conjugacy class of S_n with cycle index $(c_1, \ldots, c_n)$ is*

$$\frac{n!}{\prod c_i! i^{c_i}}.$$

PROOF: Imagine a set of parentheses where there are c_i places in i of them and the total number of places is n. There are $n!$ ways of distributing n numbers in these places. But some of the resulting permutations give the same permutation f. A cyclic permutation within a cycle does not change f nor does a permutation of the cycles of the same length. This explains the denominator in the formula above.

Note. The denominator of the lemma is also the number of permutations g for which $gfg^{-1} = f$ where f is a permutation with the given cycle index. In this way, the lemma is an example of the general fact that $|G| = |\text{stab}(x)||\text{orb}(x)|$ when G is a group acting on a set where x is an element.

R. Prove that the center of S_n consists of the unit element alone. This means that the class formula for S_n reads as follows

$$n! = 1 + \sum \frac{n!}{\prod c_i! i^{c_i}},$$

where the sum runs over all possible cycle indices $\neq (n, 0, \ldots, 0)$.

Examples

The group S_3 has three conjugacy classes with $1,3,2$ elements respectively. They are represented by, in order, the unit element, (12) and (123). The corresponding cycle indices are $(3,0,0)$, $(1,1,0)$ and $(0,0,1)$. The class formula reads simply $6 = 1 + 2 + 3$.

R. Verify the following list of conjugacy classes of S_4:

Cycle index	Number of elements	Representative
$(4,0,0,0)$	1	unit element
$(2,1,0,0)$	6	$(1)(2)(34)$
$(1,0,1,0)$	8	$(1)(234)$
$(0,2,0,0)$	3	$(12)(34)$
$(0,0,0,1)$	6	(1234)

The alternating group A_n

Consider a permutation $f(1), \ldots, f(n)$ of $1, \ldots, n$ and consider pairs j, k with $j < k$. When $f(j) > f(k)$ we say that there is a *reversal* at j with k in the second place. The number of all reversals is called the *index* $i(f)$ of f.

Example

The identity has no reversals. If f changes 1234 to 4321, there are three reversals at 1, two reversals at 2 and one reversal at 3, so that the index $i(f)$ is 6.

9.2.3 The symmetric and alternating groups

LEMMA. *If $f(j)$ and $f(k)$ change places, the index of f changes by an odd number.*

PROOF: We may assume that $j < k$. When $f(j)$ and $f(k)$ change places and $i < j$, the possible reversals at i can only change places, so that their number does not change. When $i > k$, the reversals at i are not affected at all. Hence we only have to consider the $t = k - j - 1$ places i between j and k. Suppose that there are p reversals at j with an i in the second place and q reversals at an i with k in the second place. Then there are $t - p$ non-reversals at j with an i in the second place and $t - q$ non-reversals at an i with k in the second place. Hence, when we interchange $f(j)$ and $f(k)$, the reversals and non-reversals change places so that the index changes by

$$(t - p) - p + (t - q) - q,$$

which is an even number, and by 1 or -1 coming from the change of order between $f(j)$ and $f(k)$. This proves the lemma.

The *sign* of a permutation f, denoted by $\mathrm{sgn} f$, is put to 1 when its index is even and -1 when its index is odd, in other words

$$\mathrm{sgn} f = (-1)^{i(f)}.$$

The terms even and odd are also used to the permutation itself. As we have seen above the sign of any transposition is -1.

LEMMA. *The function* $\mathrm{sgn} : S_n \to \{\pm 1\}$ *is a homomorphism.*

Note. The set $\{\pm 1\}$ is regarded as the group with two elements.

PROOF: Any permutation is a product of transpositions. By the lemma, the number of these transpositions is even when f in S_n is even and odd when f is odd. Suppose that f is a product of r transpositions and g a product of s transpositions. Then fg is a product of $r + s$ transpositions. Hence the lemma follows from the addition table for even and odd numbers and the corresponding table for 1 and -1.

It follows from the lemma that all even permutations of the symmetric group S_n (i.e. the kernel of sgn) form a group, the *alternating group* A_n.

R. Show that the alternating group has index two in the symmetric group.

R. Show that the alternating group in three variables is isomorphic to the group of rotations of an isosceles triangle.

Solvable groups

A group G is said to be *solvable* if it has a chain of subgroups

$$G = G_0 \supseteq G_1 \supseteq \cdots \supseteq G_n = \{e\},$$

where each group is a normal subgroup of the preceding one and all quotients
$$G_i/G_{i+1}$$
are abelian.

R. Show that A_3 is generated by (123) and that S_3 is solvable.

R. Show that A_4 has a normal subgroup H with elements

$$e, \ (12)(34), \ (13)(24), \ (14)(23),$$

(Klein's four group) and that this group is solvable. Hence S_4 is solvable.

R. Show that Klein's four group is isomorohic to the direct product of two cyclic groups of order 2.

Remark. It can be shown that the groups A_n with $n \geq 5$ are *simple*, i.e. they have no normal subgroups except the group itself and $\{e\}$. Hence the corresponding symmetric groups are not solvable. A famous theorem in Galois theory then implies that general algebraic equations of degree ≤ 4 over the complex numbers can be solved by extraction of roots, but that those of degree > 4 cannot. This result, due to Abel (1824), was made part of a systematic theory by Galois (1830).

Exercises

1. Let G be a group of order $2m$, m odd. Show that G has a normal subgroup of order m. (Hint. Let $s(g)$ be that sign of the permutation $x \to gx$ of G. Show that $s(g) = -1$ when g has order 2.)

2. Show that the quotient of the permutation group of four objects by Klein's four group is isomorphic to the permutation group of three objects.

3. What is the order of the permutation

$$\begin{pmatrix} 1 & 2 & 3 & 4 & 5 & 6 & 7 & 8 & 9 & 10 & 11 & 12 & 13 \\ 2 & 13 & 10 & 3 & 9 & 11 & 1 & 12 & 8 & 6 & 4 & 2 & 5 \end{pmatrix}$$

and what is its cycle index?

9.2.4 Groups of low order

Ever since the beginning of group theory, there have been efforts to determine all non-isomorphic groups of a given order. Here we will do this for small orders (the general problem is very difficult). In the sequel C_n denotes the cyclic group of order n.

R. Let G be a group such that $g^2 = e$ for all g in G. Show that G is abelian. (Hint. $ghgh = e$.)

9.2.4 Groups of low order

R. Show that a group of order 4 is isomorphic to either C_4 or $C_2 \times C_2$, which in turn is isomorphic to Klein's four group.

We are going to determine all non-abelian groups of order 6 and 8 (abelian groups were dealt with in Chapter 3).

A group G of order 6 must contain an element a of order 3 (why?). The corresponding subgroup H has index 2; hence it is normal in G. Take a b in G such that bH is the complement of H in G. From $bHbH = H$ it follows that
$$b^2 = e, \ a, \ \text{or} \ a^2.$$
The two last possibilities gives b the order 6, which is impossible if G is non-abelian. Hence $b^2 = e$. Since $bH = Hb$ we have
$$ab = ba \text{ or } ba^2.$$
In the first case, G is abelian. Hence we have the second case, i.e.
$$ab = ba^2.$$
This together with $a^3 = e$ and $b^2 = e$ determines all products of the elements of H and bH and hence of G.

R. Show that $a \to (123)$, $b \to (12)$ gives an isomorphism of G and S_3.

R. Let G be a non-abelian group of order 8. Show that it has a normal subgroup H of order 4 generated by one element a. Choose a b such that bH is the complement of H. Show that b^2 is e or a^2 and that these possibilities define two different groups, one defined by the relations
$$a^4 = b^2 = e, \ bab = a^{-1},$$
the other by
$$a^4 = e, \ a^2 = b^2, \ aba = b.$$
The last one, isomorphic to the group generated by the elements i, j, k of the quaternions, is called the *quaternion group*.

R. Which one is isomorphic to the symmetry group of a square?

R. Show that there are two non-isomorphic groups of order 10. (Hint. In a group of order 10, there is a normal cyclic subgroup H of order 5 generated by an element a and a b such that bH is the complement of H. Show that $b^2 = e$. Then show that the group is known when one knows bab (necessarily in H). Show that this gives two possibilities.)

Remark. In the last two exercises we have not, strictly speaking, proved the existence of the various groups. At this point, the reader has to rely on the word of the authors.

Use of the class formula

The class formula is extensively used in group theory. Below there are some examples and some important results.

PROPOSITION. *A group of order 15 is cyclic.*

PROOF: Let G have order 15. First we show that G is abelian. In fact, if not then the center of the group has 1,3 or 5 elements. Since the stability groups of elements outside the center have either 3 or 5 elements, the other terms of the class equation have either $15/3 = 5$ or $15/5 = 3$ elements. The stability group of an element in a conjugacy class with 3 elements has 5 elements and must be cyclic. Similarly, the stability group of an element of a conjugacy class with 5 elements has order 3.

Let us first assume that the center has just one element. The $15 = 1+3+3+3+5$ is the only way of getting 15 by adding 1 and a number of 3's and 5's. Hence there are 3 conjugacy classes with 3 elements and one with 5 elements. All the elements of the last class and no others have order 3. Hence, if an element x is in the that class so is its inverse and it follows that the class has an even number of elements, a contradiction.

Next, assume that the center has 3 elements. Then $15 = 3+3+3+3+3$ is the only way of getting 15 by adding 3's and 5's to 3. Hence all the elements outside the center have stability groups of order 5. But the center of order 3 is a subgroup of all stability groups which again is a contradiction.

Finally, assume that the center has order 5. Then $15 = 5 + 5 + 5$ is the only way of getting 15 by adding 3's and 5's to 5. Hence all stability groups of elements outside the center have order 3 and cannot contain the center. This produces a new contradiction and it follows that the center has order 15 and the group is abelian. From the main theorem on finite abelian groups in Chapter 3 it follows that G is cyclic.

R. Prove in the same way that a group of order 33 is cyclic.

R. By using for instance the class equation, prove that a group of order p^2, p a prime, is abelian.

R. Let G be a group of order p^n, p a prime. Show that the center of G has at least p elements. Show that G is solvable. (Hint. For solvability: Use induction over the order of G. Note that $G/\text{cent}(G)$ has lower order than G.)

The following theorem generalizes the fact that a subgroup of index 2 is normal.

THEOREM. *A subgroup of a group G whose index is the smallest prime dividing the order of G is normal.*

9.2.4 Groups of low order

PROOF: Let H be such a subgroup, p its index and n the order of G. Then H has n/p elements and p left cosets aH. Let T be the map from G to the permutation group of the cosets defined by

$$T(b)aH = baH.$$

The order of im$T$ then divides $p!$. An element b is in ker$T$ if and only if $baH = aH$ for all a, i.e. $ba \in aH$ or $b \in aHa^{-1}$ for all a. In other words,

$$\ker T = \cap_{a \in G} aHa^{-1} \subseteq H.$$

We have
$$n = |G| = |\ker T||\operatorname{im} T|$$

and this is only possible when im$T$ has p elements so that ker$T$ has n/p elements, i.e. as many as H. It follows that $H = \ker T$ and that H is normal.

As an application we shall show that every group whose order is the square of a prime is abelian (cf. an R above). In fact, let G be the group and p the prime. If G is not abelian, then G has an element a of order p which, by the theorem, generates a normal subgroup. Hence, for every b in G there is a k such that

$$bab^{-1} = a^k.$$

Hence
$$b^2ab^{-2} = ba^k b^{-1} = (bab^{-1})^k$$
$$= (a^k)^k = a^{k^2}.$$

Iteration gives $b^p a b^{-p} = a^{k^p}$ and $a = a^{k^p}$, since $b^p = e$. Hence we get the condition

$$k^p \equiv 1 \mod p.$$

By Fermat's theorem, the only possibility for k is $k \equiv 1 \mod p$ so that $ab = ba$ and G is abelian.

Finally, we shall give a short proof (due to McKay) of a result due to Cauchy.

R. Let G be a group of order p^n, where p is a prime and n a natural number. Suppose that G acts on a finite set X and that p does not divide $|X|$. Show that there is an x in X such that $g.x = x$ for all g in G (x is said to be a *fixed point* of G). (Hint. The set X is the disjoint union of the orbits and the length of an orbit divides the order of G.)

THEOREM. *If a prime p divides the order of a group G, then the group has an element of order p.*

PROOF: Let X be the set of all p-tuples $x = (g_1, \ldots, g_p)$ of elements of G, not all equal to e, and such that $g_1 \ldots g_p = e$. Then $|X| = |G|^{p-1} - 1$. The cyclic group C_p of order p acts on X by $c.(g_1, \ldots, g_p) = (g_2, \ldots, g_p, g_1)$, where c is a generator of C_p. By the R above, there is a fixed point of C_p in X. But such a point must have the form $(g, \ldots, g)$, which means that $g^p = e$.

Exercises
1. Let G be a group generated by the elements a and b of orders m and n respectively and suppose that

$$bab^{-1} = a^r$$

for some r. Show that
$$r^n \equiv 1 \mod m.$$

2. Let G be a group of order pq, where $p < q$ are primes and p does not divide $q - 1$.
  a) Show that G has a normal cyclic subgroup H of order q.
Suppose that H is generated by g.
  b) Show that $\mathrm{Cl}(g^k) \subseteq H$ for all k and that $|\mathrm{Cl}(g^k)| \leq q - 1$ for all k.
  c) Show that if G is *not* abelian, then $|\mathrm{Cl}(g^k)| = p$ for all k not divisible by q.
  d) Show that G *is* abelian.
  e) Show that G is cyclic.

9.2.5 Applications of group theory to combinatorics

Let $G = \{a, b, \ldots\}$ be a finite group acting on a finite set X. We have seen that $|G| = |\mathrm{orb}(x)||\mathrm{stab}(x)|$ for all x in X. Here we are going to calculate the number of orbits of X. We denote by X/G the set of orbits of G in X.

R. Show that $\mathrm{stab}(ax) = a\mathrm{stab}(x)a^{-1}$.

For a in G we let $\mathrm{fix}(a)$ denote that number of x in X such that $a.x = x$ (i.e. the number of fixed points of a).

BURNSIDE'S LEMMA. *We have*

$$|X/G| = \frac{1}{|G|} \sum_{x \in X} |\mathrm{stab}(x)| = \frac{1}{|G|} \sum_{a \in G} |\mathrm{fix}(a)|.$$

9.2.5 Applications to combinatorics

PROOF: Consider the set S of pairs (a, x) for which $a.x = x$. For a fixed a, the number of pairs (a, x) is $|\text{fix}(a)|$ and hence

$$|S| = \sum_{a \in G} |\text{fix}(a)|.$$

On the other hand, for a fixed x, the number of pairs is $|\text{stab}(x)|$, and so

$$|S| = \sum_{x \in X} |\text{stab}(x)|.$$

This last sum can be written

$$\sum_{O \in X/G} (\sum_{x \in O} |\text{stab}(x)|) = \sum_{O \in X/G} |O| \cdot \frac{|G|}{|O|} = |G||X/G|,$$

since all $\text{stab}(x)$ for x in a fixed orbit have the same number of elements. This proves the lemma.

Example

Consider the cyclic group C generated by a permutation f of n objects. Then the number of orbits of C is equal to the number of factors of f when f is written in the usual way as a product of cycles including 1-cycles. The number of orbits can also be obtained from Burnside's lemma. In fact, in the sum of the number of elements of the fixed point sets for the powers of f, the 1-cycles contribute 1 for every power of f, the 2-cycles contribute 2 for every second power etc. The total number is the number of elements of the group times the number of cycles.

Burnside's lemma leads to *Polya enumeration* when used as in the following theorem (as before we denote the set of orbits by X/G when the group G acts on the set X).

THEOREM. *Let F be the set of functions from a set $X = \{x, y, \ldots\}$ to a set U and let a group $G = \{T, S, \ldots\}$ act on X. Then G acts on F via the formula*

(1) $$(Tf)(x) = f(T^{-1}(x)).$$

Suppose that U has N elements. Then

(2) $$|F/G| = \frac{1}{|G|} \sum_{T \in G} N^{|X/\langle T \rangle|},$$

where $\langle T \rangle$ is the cyclic subgroup of G generated by T.

Note. Since every T in G permutes the elements of X, the number of orbits in X of T and STS^{-1} are the same. Hence we can sum in (2) over the conjugacy classes C of G. The result is

$$(3) \qquad |F/G| = \frac{1}{|G|} \sum_C |C| N^{|X/\langle T_C \rangle|},$$

where $\{T_C\}$ is a set of representatives for the conjugacy classes.

R. Prove that (1) defines an action of G on F.

PROOF: According to Burnside's lemma, the left side of (2) equals

$$\frac{1}{|G|} \sum_{T \in G} |\text{fix}_F(T)|.$$

Here $\text{fix}_F(T)$ is the number of functions f for which $Tf = f$, i.e.

$$f(T^{-1}(x)) = f(x)$$

for all x. Since this condition implies that $f(T^k(x)) = f(x)$ for all k, it is equivalent to the condition that f is constant on the orbits of $\langle T \rangle$ in X. Hence

$$|\text{fix}_F(T)| = N^{|X/\langle T \rangle|}$$

and this proves the theorem.

Applications

Our last theorem has a number of interesting applications.

1. *Cyclic groups*

LEMMA. *Let F be that set of functions $i \to f(i)$ from $I = \{1, 2, \ldots, n\}$ to a set with N elements and let T be the circular permutation $i \to i+1$ mod n of I. Then the action of $\langle T \rangle$ on F has*

$$\frac{1}{n} \sum_{d | n} \varphi(d) N^{n/d}$$

orbits. Here φ is Euler's function.

Note. $n = \sum_{d|n} \varphi(d)$.

9.2.5 Applications to combinatorics

Note. The number above also answers the following question: How many bracelets with n beads are there when every bead can carry N colours? A simple example: When $n = 10$, $N = 2$, there are $16(64 + 4 + 1)/10 = 112$ different bracelets.

Note. When $n = p$ is prime, (3) reads $(N^p + (p-1)N)/p$, which proves Fermat's theorem. Hence (3) extends this theorem to composite numbers.

PROOF: Adjusting the last theorem to the present situation, X is the set I and G is $\langle T \rangle$. The number of orbits in I of $\langle T^k \rangle$ is 1 when $(k, n) = 1$. In the general case, the order of T^k is $d = n/(k,n)$ and the number of orbits is $(k, n) = n/d$. Since there are $\varphi(d)$ elements of order d, the lemma follows.

2. The symmetric group

When $X = I = \{1, 2, \ldots, n\}$ and G is the permutation group of I, (3) gives the number of G when acting on functions from I to a set with N elements provided we know the number of elements in each conjugacy class.

Example

When $n = 3$, the class of the identity is the identity itself, the class of (12) has three elements and the class of (123) has two elements. Hence the formula reads

$$\frac{1}{6}(N^3 + 3N^2 + 2N)$$

in this case.

3. The symmetric group operating on switching functions

Let $B(I)$ be the set of functions $i \to x(i)$ from $I = \{1, 2, \ldots, n\}$ taking the values 0 and 1. A *switching function* is a function $x \to f(x)$ from $B(I)$ to $\{0, 1\}$. The origin of the word switching function will be explained in section 10.4. The set of switching functions will be denoted by $SW(I)$.

Since $B(I)$ has 2^n elements, there are 2^{2^n} switching functions. Via its action on $B(I)$,

$$(Tx)(i) = x(T^{-1}(i)),$$

the symmetric group acts on $SW(I)$ as follows:

$$(Tf)(x) = f(T^{-1}(x)).$$

Before sketching a way of determining the number of equivalence classes of $SW(I)$ under the action of S_I (the permutation group of I), we shall give an example.

Example

Suppose that $n = 3$ so that $B(I)$ is represented by three variables x, y, z which can take the values 0 and 1. Let the symmetric group $G = S_3$ with 6 elements permute these variables. The set of switching functions SW(I) has 256 elements. How many orbits of G are there in SW(I)? To solve this problem, we shall use Burnside's lemma in the form

$$|\text{SW}(I)/G| = \frac{1}{|G|} \sum_{T \in G} |\text{fix}(T)|,$$

where the left side is the number of orbits of G in SW(I) and fix(T) is the number of switching functions f with $Tf = f$. Since $Tf = f$ implies $STS^{-1}Sf = Sf$ and $f \to Sf$ is a bijection, $|\text{fix}(T)| = |\text{fix}(STS^{-1})|$ for all S in G. Since G has three conjugacy classes with 1,3,2 elements represented by the permutations e =the identity, (12) and (123), we have only to compute $|fix(T)|$ where T is any of these permutations. It is clear that fix(e) has $2^8 = 256$ elements. A function $f(x, y, z)$ in fix$((12))$ satisfies $f(x, y, z) = f(y, x, z)$ and is completely determined by its values on $(1, 1, z)$, $(1, 0, z)$ and $(0, 0, z)$ where z is arbitrary 0 or 1, and these values are arbitrary. This shows that $|\text{fix}((12))| = 64$. Similarly, $|\text{fix}((123))| = 16$. Hence the number of orbits in SW(I) under the action of S_3 is

$$\frac{1}{6}(256 + 3 \cdot 64 + 2 \cdot 16) = 80.$$

Let us now pass to the general case. To do this we shall reduce the computation of $|\text{fix}(T)|$ to the computation of $|B(I)/\langle T \rangle|$. This is nothing but the formula (3) in the theorem,

$$|\text{SW}(I)/G| = \frac{1}{n!} \sum_C |C| 2^{|B(I)/\langle T_C \rangle|},$$

where C runs through the conjugacy classes of S_I and $\{T_C\}$ is a set of representatives for the conjugacy classes. Now we can use the formula (3) again for the action of $\langle T \rangle$ on $B(I)$:

$$|B(I)/\langle T \rangle| = \frac{1}{|\langle T \rangle|} \sum 2^{|I/\langle T^k \rangle|}.$$

Hence our problem is reduced to the computation of $|I/\langle T^k \rangle|$ for all k and T in S_I. But the number of orbits in I under the action of the cyclic group generated by T^k is nothing but the number of cycles in the cycle decomposition of T^k. When T itself has a complicated cycle decomposition, this is not easily expressible in terms of a general formula. For instance, when $n = 6$ and $T = (1)(23)(456)$, $|I/\langle T^k \rangle|$ equals 6,3,4,5,4,3 in order for $k = 0, 1, 2, 3, 4, 5$. Hence T has $(64 + 8 + 16 + 32 + 16 + 8)/6 = 24$ orbits when acting on $B(I)$, where $I = \{1, 2, 3, 4, 5, 6\}$.

Anyway, Burnside's lemma has carried us a long way in the computation of the number of switching functions which are inequivalent under the action of the permutation group although no general formula is available.

Exercises

1. Prove that there are $(N^4 + 2N + N^2)/4$ ways of colouring the corners of a square with N colours. (Hint. Here X is the four corners of the square and the group is the rotations of the square.)

2. Let A_n be the regular n-gon with symmetry group D_n, and let n be odd. Prove that there are

$$\frac{1}{2n}(\sum_{d|n} \varphi(d) N^{n/d} + n N^{(n+1)/2})$$

ways of attaching N colours to each corner which are inequivalent under the rotations and reflections of A_n.

If n is even, prove that the same number is

$$\frac{1}{2n}(\sum_{d|n} \varphi(d) N^{n/d} + \frac{n}{2} N^{n/2}(N + 1)).$$

(Hint. The first terms in the parentheses come from the rotations, the rest from the reflections in a symmetry axis. When n is odd, each such reflection has $\frac{1}{2}(n-1) + 1$ orbits. What happens when n is even?)

Literature

There are plenty of elementary and advanced books on group theory. Burnside's lemma appeared in Burnside (1897), 165-166, although he credits Frobenius with the result. The origin of Polya enumeration is Polya (1937), where Burnside is not quoted and generating series are used extensively.

CHAPTER 10

Boolean algebra

In his book The Laws of Thought (1854) George Boole discovered that algebraic formulas and arithmetic operations can be interpreted so that they cover ordinary logic. One of his aims was to analyze the complicated statements and long lines of reasoning of philosophers. In this he was not successful since, as he says himself, their basic concepts are too vague to lend themselves to mathematical treatment. Under more precise circumstances as in the analysis of complicated combinations of simple statements he was more successful. Today, the machinery he invented, the Boolean algebras and rings, is used to analyze switching circuits. This chapter is just a simple account of finite Boolean algebras with a last section on the equivalence of Boolean functions under permutation and complementation. It does not touch the important questions of the economy of construction and complexity of circuits.

10.1 Boolean algebras and rings

Boole's basic idea was to indentify his variables $x, y, z, \ldots$ with classes of things and to give meaning to algebraic expressions like $xy, x+y, x+y+z$ etc. He identified $x, y, \ldots$ with classes of things and defined $xy = yx$ to be the class of things which are both in x and in y and $x + y$ as the class of things which are in x or in y (the inclusive addition). With these definitions he found the the distributive law

$$x(y + z) = xy + xz$$

to hold, but he had to accept that $xx = x$ and $x + x = x$ for all x. Using ∪ for union and ∩ for intersection, these formulas say that $x \cap x = x$ and $x \cup x = x$, but we shall find it convenient to stick to Boole's notation. Their familiarity and the convention of arithmetic that multiplication is performed before addition in expressions such as $xy + zu$ make them easier to read and saves a few parentheses.

We shall consider $x, y, \ldots$ to be subsets of a fixed set X, also denoted by 1. The algebra obtained in this way with xy being the intersection and

10.1 Boolean algebras and rings

$x+y$ the union of x and y, 0 the empty set and x' the complement of x in X, will be denoted by $P(X)$ and called the *set algebra* of X. The nature of computations in $P(X)$ is described in the following list of properties.

(i) Addition is commutative and associative and $x+0 = x$, $1+x = 1$ for all x

(ii) multiplication is commutative and associative and $1x = x$, $x0 = 0$ for all x

(iii) the distributive law $x(y+z) = xy + xz$ holds for all x, y, z

(iv) every x has a complement x' for which $x + x' = 1, xx' = 0$.

Note. Some of these properties follow from the others, but we shall not go into the details of this.

It is easy to prove all these properties directly by the elements of set theory, but there is also a computational thought-saving (but not space-saving) way via *characteristic* functions. To every set x we associate a function $x(t)$ from X to Z, the characteristic function of x, with the property that $x(t) = 1$ when t is in x and zero otherwise.

Example

The elements xy and $x+y$ have the characteristic functions $x(t)y(t)$ and $x(t)+y(t)-x(t)y(t)$ respectively. If $x(t), y(t), z(t)$ are denoted by, in order, a, b, c, it follows that the value of the characteristic function of $(x+y)+z$ at t is

$$(a+b-bc)+c-c(a+b-bc) = a+b+c-bc-ba-ab-ac.$$

The symmetry on the right shows that addition is associative. The other proofs are similar.

R. Verify the distributive law in this way.

Axiomatizing the properties above, we get the *Boolean algebras*.

Definition. A Boolean algebra is a set $B = \{x, y, \ldots, 1, 0, \ldots\}$ with three operations, addition $x, y \to x+y$, multiplication $x, y \to xy$ and complementation $x \to x'$ with the properties (i) to (iv) of the list above.

Note. With this definition, every $P(X)$ is a Boolean algebra.

R. Prove that $0, 1, x'$ are uniquely determined by their properties. Show that $x'' = x, 1' = 0$. (Hint. If $yx = 0, x+y = 1$, write $y = y(x+x')$.)

LEMMA. *In every Boolean algebra one has de Morgan's laws*

$$(x+y)' = x'y', \quad (xy)' = x' + y'$$

for all x and y.

PROOF: To prove the first formula, it suffices to verify that $x'y'$ has all the properties of the complement of $x + y$. We have

$$\begin{aligned} x + y + x'y' &= x(y + y') + y + x'y' \\ &= (x + x')y' + xy + y = y + y' + xy = 1 + xy = 1 \end{aligned}$$

and

$$(x+y)x'y' = xx'y' + yy'x = 0 + 0 = 0.$$

R. Show that the second law follows from the first and conversely. Prove that $x + xy = x$ for all x and y in a Boolean algebra.

Subalgebras and morphisms

A part C of a Boolean algebra B is said to be a *subalgebra* when $x + y$ and xy are in B when x and y are. A unit and zero of C are adjoined when necessary.

R. What is the subalgebra generated by 1) 0, 2) 1, 3) an element x different from the two, 4) two elements x, y different from each other and from 1 and 0?

A *Boolean polynomial* is any string which results from a finite number of Boolean operations on a finite number of elements. Repeated applications of de Morgan's laws show that in order to get the complement of a Boolean polynomial, on replaces all the variables by their complements and interchanges addition and multiplication. An interesting consequence of this procedure is that the distributive law holds when addition and multiplication change places,

$$x + yz = (x + y)(x + z).$$

R. Verify this statement.

This becomes less surprising when when the rules of computation in $P(X)$ are axiomatized with the symbols ∩ and ∪ for intersection and union, the parenthesis convention of arithmetic is not used and each operation is required to be distributive with respect to the other. The axioms are then symmetric when the two operations are interchanged.

10.2 Finite Boolean algebras

R. Prove the formula above by complementing the ordinary distributive law.

R. Prove that $x + x = x, x^2 = x$ and that $xy = y \Rightarrow x + y = x$ when x and y are any elements of a Boolean algebra. Verify the distributive law (iii) above by multiplying out the right side. Note that the cancellation law does not hold in a Boolean algebra. It does not follow that $x = y$ when $x + z = y + z$. Give an example!

R. Let n be a fixed positive integer and let B be the set of all integers ≥ 1 and $< n$. Define addition as least common multiple and multiplication as greatest common divisor. Show that B is a natural Boolean algebra when n does not have multiple factors. (Hint. When n has multiple factors, there is no complement.)

R. *Partial order.* Let $x, y, \ldots$ be elements of a Boolean algebra. We write that $x \geq y$ when $xy = y$. (The intuitive content of $x \geq y$ is that x contains y). Prove the following properties of the relation $\geq$,

$x \geq x$, (reflexivity)

$x \geq y \geq z \Rightarrow x \geq z$, (transitivity)

$x \geq y \geq x \Rightarrow x = y$, (antisymmetry).

R. Show that 'divisible by' is a partial order in the positive integers.

Boolean rings

In his book, Boole also tried to let addition be defined by the *exclusive or* meaning either x or y. In oter words, he defined the sum $x + y$ of two sets to be the set of elements belonging to x or to y but not to booth.

R. Show that in this case, the properties (i) to (iv) hold except that $1 + x = x'$. (Hint. The characteristic function of $x + y$ is now

$$x(t) + y(t) - 2x(t)y(t).$$

Direct reasoning is also possible.)

Taking (i) to (iv) with $1 + x = x$ modified to $1 + x = x'$, in particular $1+1=1'=0$, leads to an object called a *Boolean ring* with a unit. It is a ring simply because
$$x + x = 1 + x' + x = 1 + 1 = 0$$
for all x, so that addition has an inverse. The other axioms of a ring are among (i) to (iv).

R. A general Boolean ring is defined as a ring where every element equals its own square. Show that such a ring has characteristic 2. (Hint. Square $1 + x$.)

10.2 Finite Boolean algebras

We shall see in this section that every finite Boolean algebra B is isomorphic to some $P(X)$ where X is a finite set. In particular, if X has n elements, then B has 2^n elements.

Let $B = \{x, y, \ldots\}$ be a Boolean algebra. Say that x contains y when $x \geq y$, i.e. $xy = y$. Write $x > y$ when $x \geq y$ but x is not equal to y and say that an element y lies strictly between x and z when $x > y > z$.

R. Show that if one element contains two others, it contains their sum.

An element x of B is said to be an *atom* or a *minimal* element if it is not zero and there is no element strictly between x and zero. It is said to be *maximal* if it is not equal to 1 and there is no element between 1 and x.

R. Show that the complement of a minimal element is maximal and conversely.

R. Let B be the set of intervals (closed, open or half-closed) of the real line with rational endpoints. It is clear that B becomes a Boolean algebra under intersection and union. Show that this algebra does not have minimal and maximal elements.

When B is finite, however, there are maximal and minimal elements. In fact, every $x > 0$ contains atom. To see this, we remark that if x is not an atom, there is an y such that $x > y > 0$. If y is not an atom, repeat this process which, since B has a finite number of elements, must finish with an atom contained in x.

R. Prove that the product of two atoms is zero unless they are identical.

LEMMA. *If $x > y$, then x contains an atom not contained in y.*

Proof. That $x > y$ means that xy' is not zero. Hence this element contains an atom z, $xy'z = z$. Hence $zy = xyy'z = 0$ but, obviously, xz is not zero. This finishes the proof and also most of the proof of

THEOREM. *An element of a finite Boolean algebra is the sum of the atoms it contains and the product of the maximal elements which contain it. Two sums of atoms are the same if and only if their atoms are the same.*

PROOF: The first part of the theorem follows from the lemma and the second one by complementation. To prove the third part, let

$$x_1 + \cdots + x_n = y_1 + \cdots + y_m$$

be two sums of atoms. Multiplying both sides by one atom of the left side reproduces it. Hence it must occur also on the other side. This proves that all atoms of the right side are contained in the left side and conversely. The proof is finished.

R. Let B be a finite Boolean algebra, X the set of its atoms and $f : B \to P(X)$ a map which maps an element x of B to the set of its atoms. Prove that f is a morphism in the sense that it is bijective and turns the operations of B into the corresponding set operations. (Hint. Use the theorem.)

When x is an element of a Boolean algebra and $y_1, \ldots, y_n$ are all the atoms it contains, we have

$$x = y_1 + \cdots + y_n,$$

a formula which in the applications is called the *disjunctive* normal form of x. Similarly, if $z_1, \ldots, z_n$ are the maximal elements containing x, we have the *conjunctive* normal form

$$x = z_1 \ldots z_n.$$

R. Let E be a Boolean algebra with elements 0 and 1. Show that all functions from a set X to E constitute a Boolean algebra $F(X, E)$ with the natural operations. Show that $F(X, E)$ is isomorphic to $P(X)$. Show that the atoms of $F(X, E)$ are the functions $f_y(x)$ which take the value 1 when $y = x$ and 0 otherwise.

Boolean functions

Let E be a Boolean algebra with the elements 0 and 1 and let $X = E^n$ be the product of n copies of E with elements $x = (x_1, \ldots, x_n)$ and let $F = F(X, E)$ be the Boolean algebra of functions $f(x)$ from X to E. The elements of such a Boolean algebra are called *Boolean functions*. In this way, Boolean polynomials in $x_1, \ldots, x_n$ are elements of F. Two Boolean polynomials are said to be equal if they are equal as functions. This does not mean that they are the same as algebraic expressions. To get uniqueness, we have to write them in normal form,, for instance in disjunctive form.

Example

A Boolean algebra generated by two elements x, y not equal to each other or to 0 and 1 has four atoms, viz. $xy, x'y, xy', x'y'$. The element 1 is the sum of all of them, x equals $xy + xy'$, $x + y$ equals $x(y + y') + y(x + x') = xy + xy' + yx'$ and so on.

R. Write the following Boolean functions of x, y, z in disjunctive normal form: a) $x + x'y$, b) $xy' + xz + xy$, c) $xyz + xy + xz$.

Applications to circuits

A real world electrical circuit consists of wires connecting one or several voltage input gates with an output gate and it has a number of switches which disconnect or connect a wire. If we are only interested in presence or absence of voltage at the output gate, we can make an efficient mathematical model of the circuit by introducing a Boolean variable for each switch which equals 0 when the switch is open (does not let current through) and 1 when it is closed. If $x_1, \ldots, x_n$ correspond in this way to the switches the observed voltage at the output is a Boolean function $f(x_1, \ldots, x_n)$ from E^n to E. Conversely, to every such function there is a corresponding circuit. For example, a product $x_1 x_2$ is realized by a series coupling and the sum $x_1 + x_2$ as a parallell coupling according to the figure below. To construct a circuit for a given Boolean function f of a finite number of variables which generate a Boolean algebra, we only have to write it in disjunctive normal form and then realize the atoms, which are products of the generators and their complements as series couplings and then join them through parallel couplings. Practically, this is not quite perfect since in this way a gate corresponding to one variable may be along way from a gate corresponding to its complement. Anyway, Boolean algebra is a routine tool in the analysis and construction of circuits.

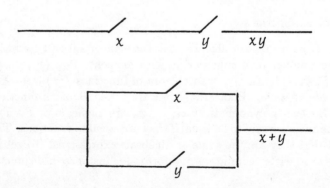

Exercises
1. Show that every non-empty subset $a > x > b$ of a Boolean algebra is a subalgebra.
2. How many elements are there in $F(E, E^n)$?
3. Draw a circuit for
$$x_1(x_2(x_3 + x_4) + x_3(x_5 + x_6))$$
and simplify it.
4. Let a be a non-zero element of a Boolean ring A with a unit 1. Show that R is the direct sum of aR and $(1+a)R$. Show that if R is finite, it is the direct sum of a finite number of Boolean rings isomorphic to Z_2.

10.3 Equivalence classes of switching functions

Consider a circuit S with n switches $I = \{1, \ldots, n\}$. Let $B(I)$ be the set of n-tuples $x = (x(1), \ldots, x(n))$ where $x(i) = 0$ or 1 denotes the on and off position of the corresponding switch. The output voltage,
$$F(x) = F(x(1), \ldots, x(n)),$$
observed at the output gate and also assumed to be 0 or 1, is then a Boolean function which in this case is also called a switching function. The set of switching functions on a set I of switches will be denoted by SW(I).

In section 9.2.5 we have determined, in principle, the equivalence classes of switching functions under the action induced by permutations $j \to T(j)$ of switches and actions $(Tx)(i) = x(T^{-1}(i))$ on their positions. To this we shall now add permanent reversions of some of the switches. The corresponding operation on Boolean functions is to *complement* certain variables. i.e. replace certain variables $x(i)$ by $x(i)'$ where $0' = 1, 1' = 0$. Our problem is compute the minimal number of switching devices which, by these modifications, generate all Boolean functions of n variables. Before going into the general situation, we give an example.

R. Prove that the 6 Boolean polynomials $f(x, y) = 1, 0, xy, x, x + y, 1 + xy$ generate all 15 Boolean functions of two variables under the action of permutations and complementations of the variables. (Hint. If $f(x, y) = x$ and $x + y$ are chosen, they generate, in order, all Boolean functions with no, all, one, two and three ones in their ranges. Prove also that the number 6 is minimal.)

To treat the general situation, we shall make the values 0,1 of the variables $x(i)$ elements of the Boolean ring Z_2. In this way the operation
$$x \to x + y = (x(1) + y(1), \ldots, x(n) + y(n))$$

expresses complementation of $x(i)$ when $y(i) = 1$ and no complementation when $y(i) = 0$. This gives a simple way of treating the group $G(I)$ generated by of permutations and complementations of the space $B(I)$ which now is a Boolean ring. This group acts in turn on Boolean functions via the formulas

$$F(x) \to F(T(x)), \quad F(x) \to F(x+y).$$

(For convenience we do not introduce separate notations for the functions after the arrows.) In terms of the group $G(I)$, our problem is to compute the number of equivalence classes of Boolean functions under the action of $G(I)$. Later we shall also consider the corresponding problem when also complementation $F(x) \to F(x)'$ of their values is included. This corresponds to a reversal of an output switch.

Our computation will use Burnside's lemma and we therefore have to begin by a study of the group $G(I)$ itself.

The group $G(I)$ of permutations of complementations of $B(I)$

LEMMA. *The permutations and complementations generate a group $G(I)$ of bijections of $B(I)$. Its order is $n!2^n$ and its elements $x \to f(x)$ have the form*

(1) $$f(x) = Tx + y, \quad f^{-1}(x) = T^{-1}(x+y)$$

where y and the permutation T are uniquely determined by f. The product of f by g, given by $g(x) = Sx + y$, is

(2) $$(fg)(x) = TSx + Tz + y,$$

and the conjugate gfg^{-1} of f by g is

(3) $$gfg^{-1}(x) = STS^{-1}(x) + z + STS^{-1}(z).$$

The group of complementations form a normal subgroup of $G(I)$ of order 2^n.

Note. The formulas above show the complete analogy between $G(I)$ and, for instance, the group of rotations and translations of Euclidean space.

PROOF: The formula (2), which is the result of direct computation, shows that every element of $G(I)$ has the form (1) where, obviously, $f(x) = x$ for all x if and only if $y = 0$ and T is the identity. Hence the order of $G(I)$ is $n!2^n$. The formula (3), also the result of direct computation, shows that the group of complementations $x \to x + y$, form a normal subgroup, obviously of order 2^n. This finishes the proof.

10.3 Equivalence classes of switching functions

R. Prove that $G(I)$ acts transitively on $B(I)$, i.e. given any pair of elements x, y of $B(I)$, there is an f in $G(I)$ such that $f(x) = y$.

The cycle structure of $G(I)$

Let $(c(1), \ldots, c(n))$ be the cycle structure of the permutation T, i.e. $c(k)$ is the number of orbits of T with k elements. As shown in the preceding chapter, the cycle index is a complete invariant of the class of T in $S(I)$, the group of permutations of $I = \{1, 2, \ldots, n\}$. More precisely, all elements of the conjugacy class of T, i.e. permutations of the form STS^{-1} with S an arbitrary permutation, and only these have the same cycle index as T. We shall now construct cycle indices of elements $x \to f(x) = Tx + y$ of $G(I)$.

Definition. The cycle index of f equals

$$(c(1), \ldots, c(n))[e(1), \ldots, e(n)]$$

where $(c(1), \ldots, c(n))$ is the cycle index and $e(k)$ is the number of cycles C of T of length k for which $\sum_C y(j) = 1$ (in B), i.e. $y(j) = 1$ an odd number of times for j in C.

Note. The parentheses [.] are used only to set off the second part of the cycle index from the first part.

THEOREM. *The cycle index defined above characterizes completely the conjugacy class of f in $G(I)$.*

PROOF: The kth iterate of an element $f(x) = Tx + y$ of $G(I)$ is

$$f^{(k)}(x) = T^k x + y + Ty + \cdots + T^{k-1} y,$$

which means that

(4) $\qquad (f^{(k)} x)(j) = x(T^{-k}(j)) + y(j) + \cdots + y(T^{(k-1)}(j)).$

If $I = C \cup D \cup \ldots$ is a partition of I into cycles of T, and $B(C)$ denotes the set of functions $j \to x(j) = 0$ or 1 with j in C, $B(I)$ is the direct sum

$$B(I) = B(C) \oplus B(C) \oplus \ldots.$$

It follows from (4) that f acts separately on each term and that if $h = gfg^{-1}$, then h acts on $g(B(C)) = B(S^{-1}C)$ precisely as f acts on $B(C)$. Hence it suffices to study the action of f on a general cycle C of T. If $k = |C|$ is the number of elements of C, then (4) shows that $(f^{(k)}(x))(j) = (f(x))(j)$ if and only if $\sum_C y(j) = 0$. Otherwise, the right side is $(f(x)(j))'$ and the order of f, restricted to C, is $2k$. Next we shall see that the cycle indices of f and $h = gfg^{-1}$ are the same. The cycle C for T corresponds to the cycle SC for STS^{-1}, and to verify our statement, (3) shows that we only have to prove that the

$$\sum_C y(i) = \sum_{SC} y(S^{-1}i) + \sum_{SC} (z(i) + z(ST^{-1}S^{-1}(i))).$$

But this is clear since $ST^{-1}S^{-1}$ is a bijection of SC so that the last sum on the right vanishes.

Conversely, suppose that f and $h(x) = Rx + u$ have the same cycle index. Then their first parts are the same so that, by the theory of the symmetric group (see section 9.2.3), there is a permutation S such that $SRS^{-1} = T$. Hence, if $g(x) = S(x) + z$, the formula (3) shows that

$$(ghg^{-1})(x) = Tx + Su + z + Tz$$

and we have to solve the equation $y = Su + z + Tz$ for z. It suffices to do this for every cycle C of T on I. We are then are free to choose our notations so that $C = (1, \ldots, k)$ and $T(j) \equiv j + 1 \mod k$. In this situation, the equation says that

$$z(j) + z(j+1) = y(j) + u(S^{-1}(j)).$$

By hypothesis, the sum over j on the right side vanishes and hence it suffices to prove that a system of equations

$$z(j) + z(j+1) = v(j)$$

where z and v are functions from Z_k to B is solvable when $\sum v(j) = 0$. But this is clear. It suffices to put $z(j) = v(1) + \cdots + v(j-1)$ when $j > 1$ and $z(1) = 0$. This gives two values for $z(k)$, namely $v(k)$ and $v(1) + \cdots + v(k)$ and they are equal. This finishes the proof of the theorem.

Cycle notation for the elements of $G(I)$

A permutation T can be written as $CD \ldots$ where $C, D, \ldots$ are the orbits or cycles of T when acting on I, $C = (j, Tj, \ldots)$. A similar notation can be used for the bijections

$$x \to f(x) = Tx + y$$

of $B(I)$. When $y(k) = 1$, we write k' instead of k in the cycle decomposition. The meaning of this is that

$$(f(x))(k) = x(T^{-1}(k)), \quad (f(x))(j) = x(T^{-1}(j)) \quad \text{when} \quad j \neq k.$$

10.3 Equivalence classes of switching functions

Examples

$$((1'))(x) = (x(1)', x(2), \ldots),$$
$$((1'2))(x) = (x(2)', x(1), \ldots),$$
$$((1'2'))(x) = (x(2)', x(1)', \ldots)$$

etc. Orbits with an odd number of elements are said to be odd.

By the previous lemma, every element of $G(I)$ is conjugate to an element f whose cycle decomposition as above has at most one primed integer in every cycle of its permutation.

The table below lists all cycle indices of $G(I)$ when I has three elements, typical elements of the corresponding classes (typel), the number of elements of the corresponding conjugacy class (|els|), the number of orbits in $B(I)$ of a member of the class (|$B(I)$orb|) and the number of elements with at least one odd orbit (|oddorb|).

Table of the action of $G(I)$ on $B(I)$ when $|I| = 3$.

cycle	index	typel	|els|	|$B(I)$orb|	|oddorb|
$(3,0,0)$	$[0,0,1]$	$(1)(2)(3)$	1	8	8
$-$	$[1,0,0]$	$(1')(2)(3)$	3	4	0
$-$	$[2,0,0]$	$(1')(2')(3)$	3	4	0
$-$	$[3,0,0]$	$(1')(2')(3')$	1	4	0
$(1,1,0)$	$[0,0,0]$	$(1)(23)$	6	6	6
$-$	$[1,0,0]$	$(1')(23)$	6	2	0
$-$	$[0,1,0]$	$(1)(2'3)$	6	4	0
$-$	$[1,1,0]$	$(1')(2'3)$	6	2	0
$(0,0,1)$	$[0,0,0]$	(123)	8	4	8
$-$	$[0,0,1]$	$(1'23)$	8	2	0

The number of orbits on $B(I)$ have been computed by hand. For instance, the following actions

$$(1'23) : (000) \to (110) \to (111) \to (001) \to (001) \to (000)$$

and

$$(1'23) : (010) \to (101) \to (010)$$

prove that the action of $(1'23)$ has two orbits on $B(I)$ with 6 and 2 elements respectively.

Our table permits an immediate computation of the number of orbits of the action of $G(I)$ on SW(I) when I has three switches. Using the formula

(2) of section (9.2.5) we get

(4) $$|SW(I)/G(I)| = (\sum_C 2^{|\mathrm{orb} f|})/48$$

where the sum runs over all conjugacy classes C of $G(I)$, f is any element of C and orb$f$ is the set of orbits of f acting on $B(I)$. An insertion of the numbers of the table shows that there are 1056/48=22 inequivalent switching functions. Note that the corresponding number under equivalence only under permutation was 80 (see the end of section 9.2.5).

R. Write down the corresponding table for $n = 2$ and verify that that there are 6 equivalence classes in this case.

R. Prove that the Boolean polynomials $f(x,y) = 1, xy, x+y$ generate all 16 Boolean functions of two variables under permutation and complementation of the variables and complementation of the output. (Hint. The preceding exercise.)

Now consider actions

$$F(x) \to ((f,a)F)(x) = (f^{-1}(x)) + a$$

on switching functions where f is in $G(I)$ and a is in B. The notation is meant to indicate that the pair (f,a) operates on the switching function F. It is obvious that $(f,0)(g,0) = (fg,0), (f,0)(g,1) = (fg,1), (f,1)(g,1) = (fg,0)$, so that all these actions form a group $H(I)$ with $2|G(I)| = n!2^{n+1}$ elements. To use Burnside's lemma for this group, we have to compute $|\mathrm{fix}(f,a)|$ for all (f,a).

LEMMA. *The number of elements of* fix$(f,0)$ *is* $2^{|\mathrm{orb} f|}$ *where* orb *f means the set of orbits of f in $G(I)$. The set* fix$(f,1)$ *is empty unless all orbits of f in $G(I)$ have an even number of elements and in this cases it has* $2^{|\mathrm{orb} f|}$ *elements. The number of elements in each case depends only on the class of f in $G(I)$.*

PROOF: The first assertion is obvious since f only has two values on each orbit. To prove the second one, consider the kth iteration of the action above with $a = 1$,

$$F(x) \to F(f^{(k)}(x)) + k.$$

It follows that for fix$(f,1)$ to be not empty, k has to vanish in B when $f^{(k)}(x) = x$ and this for all x. This is the case if and only if all the cycles of f have an even number of elements. Hence the second assertion follows. The third one is a consequence of the invariance of cycle indices under conjugation.

10.3 Equivalence classes of switching functions

The lemma shows how to compute the number of equivalence classes of $SW(I)$ under the action of $H(I)$. It follows from the table above that when $|I| = 3$, this number is

$$22 - (\sum_C 2^{|\mathrm{orb} f|}/96)$$

where now C runs over the conjugacy classes containing elements with at least one odd orbit and f is a representative of this class. According to the table above, the sum within parenthesis is

$$2^8 + 6 \cdot 2^6 + 8 \cdot 2^4 = 768 = 8 \times 96.$$

It follows that $SW(I)$ has $22 - 8 = 14$ inequivalent functions under the action of $H(I)$ when $|I| = 3$.

The general case

For $I = \{1, 2, \ldots, n\}$ with n arbitrary, one follows the path indicated in the case of three switches. In the first place one needs to know the number of elements of a conjugacy class C of $G(I)$ with a given cycle index

$$(c(1), \ldots, c(n))[e(1), \ldots, e(n)].$$

This number is given by the formula

$$|C| = n! \prod_i (2^{(i-1)c(i)}/i^{c(i)}(c(i) - e(i))!),$$

but the orbits of $B(I)$ under a bijection $f(x)$ have still to be found by hand. The case $|I| = 4$ attracted interest in the sixties (Harrison 1965). He found 402 equivalence classes without complementation and 222 in the general case and he also listed 222 devices which generate all 65536 Boolean functions of four variables under permutations and complementations of the variables and the output.

A reader who does not care for the practical use of switching functions with three of four variables may perhaps still appreciate the group theory used and the way it has made complicated bookkeeping problems tractable.

Literature

The cycle structure of the group of permutations and complementations of a finite Boolean algebra is due to A. Young (1929). The application to switching functions presented here is a simplified version of Harrison (1965).

CHAPTER 11

Monoids, automata, languages

The origin of the modern theory of automata was a paper by Alan Turing in 1936. His automaton, the Turing machine, was designed to imitate step by step computing. The machine had a finite number of internal states and a printing head that could print zeros and ones on an infinite tape. In each step, the machine jumped from one state to another and the tape was moved left or not at all and printed 0 or 1. Which of these possibilities that ocurred was made to depend on the state of the machine before the step and the last printed symbol.

The machine started from an initial state and an empty tape and the sequence of zeros and ones to the right of the original square was interpreted as a number between 0 and 1. The question that Turing asked was: can every number between 0 and 1 be produced by this machine if it is allowed to work indefinitely? Using a resoning analogous to Cantor's proof that the real numbers are uncountable, Turing gave a negative answer to his question.

One of Turing's achievements in this paper was to use the notion of an automaton to define computability. This gave rise to a new branch of the theory of formal languages, namely the characterization of the formal languages which are produced by automata. This branch is now part of theoretical computer science. The aim of this chapter is to demonstrate the use of matrix calculus to prove one of the first results of the theory, namely Kleene's characterization of rational or regular languages as those produced by automata of a certain simple kind.

11.1 Matrices with elements in a non-commutative algebra

The part of classical algebra to be used in this chapter is the calculus of matrices with coefficients in an algebra, defined to be a set $R = \{a, b, c, \dots\}$ with associative addition and multiplication where the right and left distributive laws hold,

$$a(b+c) = ab + ac, \quad (b+c)a = ba + ca.$$

Multiplication is not required to be commutative. An algebra in this sense differs from a ring in that subtraction is not required.

A matrix with elements in R is a rectangular scheme A with elements in R is written as
$$A = (a_{jk}),$$
where $j = 1, \ldots, p$ and $k = 1, \ldots, q$ and the pq elements a_{jk} are in R. A matrix of this form is said to be of type $p \times q$. Precisely as for matrices with with numerical elements, addition of two matrices of the same type is defined by
$$A + B = (a_{jk} + b_{jk}).$$
A product AB is only defined when A has type $p \times q$ and B has type $q \times r$ and then it is given by
$$AB = (\sum_{k=1}^{q} a_{jk} b_{kl})$$
where $j = 1, \ldots p$, $k = 1, \ldots, r$. The order in the products is important. One verifies immediately that addition and multiplication are associative and that the distributive law holds from both sides when every term is defined.

An important tool in matrix theory is the possibility to use block matrices. Let $N = \{N_1, \ldots, N_p\}$ be a partition of $\{1, \ldots, n\}$ into p blocks of integers which follow each other in the natural order. Let n_r be the number of elements of N_r so that $n_1 + \cdots + n_p = n$, Any matrix of type $n \times n$ can be considered as a matrix (A_{jk}) of blocks A_{jk} of $n_j \times n_k$ matrices defined by
$$A_{jk} = (a_{pq}) \text{ with } p \text{ in } N_j, \ q \text{ in } N_k.$$
The point of this is that if B_{jk} are the corresponding blocks of another $n \times n$ matrix, then the blocks of the $n \times n$ matrix $C = AB$ are
$$C_{jk} = (\sum_{i=1}^{p} A_{ji} B_{ik}),$$
precisely as for general matrix multiplication. The proof is immediate. The elements of C_{jk} are simply
$$\sum_{t=1}^{n} a_{pt} b_{tq}$$
where p is in N_j and q is in N_k. Cutting up the sum into r pieces with t in $N_1, N_2, \ldots, N_r$ proves the desired result.

11.2 Monoids and languages

11 Monoids, automata, languages

A *monoid* M is simply a set of elements $u, v, w, \ldots$ with an associative multiplication, not necessarily commutative. It may or may not have a unit, i.e. an element e with the property that $eu = ue = u$ for every u in M. A monoid with a unit differs from a group by not having inverses of all its elements. A *submonoid* of M is any part of M which is a monoid with the same multiplication. Any subset N of a monoid M generates a submonoid N^* consisting of all products of elements of N.

R. Let M be a monoid with a unit e. Let u be an element of M not equal to e and put $N = u^* \cup e$ with $u^0 = e$. Show that unless all powers of u are separate, there is a least natural number n for which there exists a number $m < n$ such that $u^m = u^n$. Show that, in the latter case, all elements of N are powers of u with exponent $< n$. When is N a group?

Our main interest will be the free monoid Σ^* generated by a finite set Σ just by forming all finite products $uvwvuw\ldots$ where $u, v, w, \ldots$ are elements of Σ. The elements of Σ will be called letters, those of Σ^* will be called words and Σ itself an alphabet. The word free means that two words are considered to be the same only when they consist of the same letters in the same order. The role of a unit is played by the empty word ϵ with the property that $\epsilon w = w\epsilon = w$ for every word. Subsets of $\Sigma^* \cup \epsilon$ will be called *languages*.

Example

All written languages are languages in the abstract sense, in particular the programming languages of computer science.

Formal series

It is convenient to think of a language L as a formal series

$$L = \sum (L, w)w$$

where w runs through the words of Σ^* and $L(w)$ is zero or 1 according as w is in L or not. Here 0 and 1 should be the elements of the Boolean algebra $B = \{0, 1\}$ with the structure

$$0 + 0 = 0, \quad 1 + 0 = 0 + 1 = 1, \quad , 1 + 1 = 1, \quad 00 = 0, 10 = 01 = 0, 11 = 1.$$

The terms of a series L can be taken in any order such that every partial sum with a sufficient number of terms contains any given word of the language. In the sequel we shall identify a language and its series, also called formal power series.

The sums and product of the formal series U and V are defined by

$$U + V = \sum ((U, w) + (V, w))w, \quad UV = \sum (U, s)(V, t)w \text{ for } w = st.$$

11.2 Monoids and languages

In this way the words of $U+V$ is the union of the words of U and V and the words of UV are those of the form st where s is a word of U and t is a word of V. With these definitions, all languages with the alphabet Σ form an algebra $B\langle\langle\Sigma\rangle\rangle$ over B with addition and multiplication as above. It is easy to verify that both the left and right distributive laws are satisfied. The algebra has a zero, namely the zero language, (all $(L,w)=0$), and its unit is ϵ where $(\epsilon,w)=0$ except that $(\epsilon,\epsilon)=\epsilon$. Note that since addition in the Boolean algebra B does not have an inverse, the same is true of $B\langle\langle\Sigma\rangle\rangle$. The elements of $B\langle\langle\Sigma\rangle\rangle$ with a finite number of terms will be called *polynomials*. They generate an algebra which will be denoted by $B\langle\Sigma\rangle$. The corresponding languages are finite.

R. Show that $B\langle\langle\Sigma\rangle\rangle$ is an idempotent algebra in the sense that $U+U=U$ for every U in the algebra.

R. By our identification of languages with power series, we have

$$U^* = \epsilon + U + U^2 + \dots$$

Show that $(U^*)^* = U^*$ by taking the limit of U^*U^n for $n \to \infty$.

Note. Later we shall also use the notation

$$U^\wedge = U + U^2 + U^3 + \dots$$

The right side of this formula will be called the *pseudoinverse* of U.

Note. When B is replaced by a commutative ring A, the algebra

$$A\langle\langle\Sigma\rangle\rangle$$

of formal series in Σ with coefficients in A. is a ring since the series $-L$ with coefficients $-(L,w)$ is the inverse of L.

Polynomials. Rational languages

All languages generated by a finite number of additions, multiplications and the operation $U \to U^*$ starting from the empty language and the letters of a fixed alphabet Σ are said to be *rational*.

Note. The use of the word rational comes from an analogy with the rational power series of complex rational functions which vanish at the origin. In fact, these rational functions all have the form

$$f(z) = p(z)/(1-q(z)) = p(z)q(z)^*$$

when $p(0) = q(0) = 0$. It is immediately verified that they form a ring and that f^* has the same form as f. The same result holds for polynomials in one variable with coefficients in a ring. The fact that there are more complex power series than the rational ones above is an indication that the same should hold for the general case, namely that the rational languages form just a small subset of all languages.

Note. When B is replaced by an arbitrary algebra or ring, U^* exists when $(U, \epsilon) = 0$, but may not exist otherwise, for instance not when Σ has one letter a and $U = \epsilon + a$. In fact, $U^n = \epsilon + na$. In these cases, the rational languages are defined as above but with U^* replaced by $U^\wedge$.

Example
If Σ has the letters a, b, c, the following series are rational,

$$(a+b)^*, (a+bc^*)^*c^*, (abc)^*(c+a^*cbca^*)^*.$$

Expressions like these are not unique in any way. We have, for instance,

$$(a+b)^* = a^*(ba^*)^*.$$

In fact, every element of $(a+b)^*$ is a product of a power of a and products bc where c is a power of a, 1 included and every such product occurs.

In non-technical terms we may describe rational languages as follows. When writing the words of a rational language, one is allowed to write any finite sequence of words, repeat any part which one has just written any finite number of times and then start again.

Exercise
At this point we need a proof that not all languages are rational. To see this, consider languages generated by a single letter x. Define a gap in a power series (language)

$$r = \sum_{n \geq 0} a_n x^n, \quad n = 0, 1, \ldots$$

to be a sequence of vanishing letters between two non-vanishing ones. Prove that a rational language does not have single power series with gaps of arbitrary length. (Hint. This is clear for polynomials. Prove it for sums, products and for the stars. Hence there are plenty of languages which are not rational.)

11.3 Automata and rational languages

Generally speaking, an automaton is a machine with a number of buttons and a number of states. Pushing down one button makes the machine

11.3 Automata and rational languages

change (or not change) its state. At the same time the machine may or may not print something. We shall ignore the printing and restrict ourselves to the case when the automaton has a finite number of buttons, called inputs, and a finite number of states.

Our description of an automaton will be limited to listing its states and, for every change of state, the inputs which cause this change. A convenient way of doing this is to introduce its *transition matrix*. It is a square matrix M labelled by the states $p, q, \ldots$ in such a way that the entry $M(p,q)$ is the subset of inputs I which, when applied to the automaton in the state p, puts it into the state q. We shall write this as

$$pM(p,q) = q,$$

making the states in $M(p,q)$ operate from the right. This is of course a rather theoretical situation but enough for the next item of interest, namely the language accepted by the automaton. This language has as letters Σ the entries of the transition matrix with the convention that an empty entry is replaced by ϵ.

Example
Let the states be p and q and let the transition matrix be

$$M = \begin{pmatrix} x & y \\ 0 & u \end{pmatrix}$$

which means that $xp = p, yp = q, uq = q$. It is also possible to visualize this automaton by its graph. It has two points p, q and x, y, u are represented by directed lines connecting the two points or a point to itself.

R. Draw this graph.

Every automaton accepts a number of languages, namely the set of words with letters from its transition matrix which are generated when successions of inputs which, when are applied to a given state of the automaton, carry it into another given state. For the automaton above, this procedure gives four languages, namely

p to $p : x^*$, p to $q : x^*yu^*$, q to $p : 0$, q to $q : u^*$.

This can also be expressed in terms of the transition matrix M. We observe that the elements of M^2 are the words with two letters taking a given state into a given state and similarly for the other powers. Hence the elements of

$$M^* = \sum M^k, \quad k = 0, 1, 2, \ldots$$

are precisely the four langauages of the preceding formula. We note that the four languages above and hence also their sum is rational. This fact is general and will be proved presently.

THEOREM. *The languages accepted by a finite automaton with a rational transition matrix are rational.*

PROOF: Suppose that the transition matrix M of the automaton is a $n \times n$ matrix whose elements are rational series in $B\langle\langle\Sigma\rangle\rangle$ with Σ a finite alphabet. Write M as

$$M = \begin{pmatrix} M(1,1) & M(1,2) \\ M(2,1) & M(2,2) \end{pmatrix}$$

where $M(1,1)$ and $M(2,2)$ are square matrices of orders 1 and $n-1$ respectively and $M(1,2)$ and $M(2,1)$ have the types $1 \times (n-1)$ and $(n-1) \times 1$ respectively. We shall see that the corresponding division of M^* has the elements

$$\begin{aligned} M^*(1,1) &= (M(1,1) + M(1,2)M(2,2)^*M(2,1))^* \\ M^*(1,2) &= M(1,1)^*M(1,2)(M(2,2)^* + M(2,1)M(2,2)^*M(1,2))^* \\ M^*(2,1) &= M(2,2)^*M(2,1)(M(1,1) + M(1,2)M(2,2)^*M(2,1))^* \\ M^*(2,2) &= (M(2,2) + M(2,1)M(1,1)^*M(2,1))^* \end{aligned}$$

To see this let us first lump the states corresponding to the $n-1$ last rows and columns of M to a state 2 and let state 1 be the remaining state.

To prove the first formula, we simply note that any word with letters $M(j,k)$ which maps the state 1 into itself is a product of elements where either no $M(1,2)$ and $M(2,1)$ appear and only $M(1,1)$ or else $M(1,2)$ and $M(2,1)$ appear together in this order with sequences of $M(2,2)$ between them and sequences of $M(1,1)$ between them when they are taken in opposite order. This proves that $M^*(1,1)$ and, by symmetry, also $M^*(2,2)$ has the form stated.

To prove the second formula, note that a word taking state 1 to state 2 starts with a sequence of $M(1,1)$ followed by $M(1,2)$ which in turn is followed by any sequence mapping state 2 into state 2. Hence the form of $M^*(1,2)$ follows and, by symmetry, also that of $M^*(2,1)$. We can now prove the theorem. If the languages accepted by $M(2,2)$ are rational, the lemma shows that those accepted by M are also rational. Hence the theorem follows by induction over the number of states.

11.4 Every rational language is accepted by a finite automaton

The following conventions are followed in the literature. From the states s of a finite automaton, one state s, called the initial state, is singled out and also a subset F of S whose states are said to be final. The words of the language accepted by the automaton is defined as the set of words mapping s into a state in F. We can then arrange for s to label the first row of the transition matrix A and for the elements of F to label the columns. The

11.4 Every rational language is accepted by a finite automaton

language (or powers series) accepted by the automaton is then simply the sum of the elements of the first row of A^*. The following theorem is due to Kleene (1956)

Theorem. *Every rational language is accepted by a finite automaton.*

Note. By suitable modifications in the proof below, this result holds also for formal power series with coefficients in an arbitrary commutative algebra (The Kleene-Schützenberger theorem).

Proof: Before proceding to the proof proper, we need som notation. The first row of an $n \times n$ matrix A will be denoted by

$$A_1 = (A_{11}, \ldots, A_{1n}).$$

The sum of the elements of a matrix C will be denoted by $|C|$. The sum of the elements of A_1^*, denoted by $\|A\|$, is called the principal language generated by A.

We shall first show that letters, sums, products and pseudoinverses of languages with a fixed alphabet are accepted by automata.

To start with letters, let L be a letter and consider the automaton

$$A = \begin{pmatrix} \epsilon & L \\ 0 & \epsilon \end{pmatrix}.$$

The first row of $A^* = A$ is the empty word and L and hence A accepts L. Next, consider sums. Let A and B be generating matrices of two automata and consider the matrix

$$C = \begin{pmatrix} \epsilon & A_1 & B_1 \\ 0 & A & 0 \\ 0 & 0 & B \end{pmatrix}.$$

Reasoning as in the proof of the theorem of the last section, it is easy to see that

$$C^* = \begin{pmatrix} \epsilon & A_1 A^* & B_1 B^* \\ 0 & A^* & 0 \\ 0 & 0 & B^* \end{pmatrix}.$$

The first row of this matrix is $(\epsilon, A_1^{\wedge}, B_1^{\wedge})$. Hence the language accepted by C is the sum of those accepted by A and B.

Next, consider products. Let A and B be square matrices of orders p and q and consider the matrix

$$C = \begin{pmatrix} A & PQ \\ 0 & B \end{pmatrix}.$$

Here P is the $p \times q$ matrix whose first column has all its elements equal to ϵ while all other elements are 0 and Q is a $q \times p$ matrix with ϵ in the upper left corner and 0 in other places. The formula for M^* of the last section shows that

$$C^* = \begin{pmatrix} A^* & A^*PQB^* \\ 0 & B^* \end{pmatrix}.$$

Here A^*P is a $p \times q$ matrix whose first row consists of the sums of the elements of the rows of A^* while the others vanish and the first row of QB^* is that of B^* while the others vanish. Hence the first row of C^* consists of the first row of QB^* multiplied on the left by $||A||$. Hence $||A||\,||B||$ is the language generated by C.

Finally, we have to show that if a language L is generated by an automaton, so is L^*. Too see this, observe first that if the matrices P and Q above refer to square matrices, then

$$|A_1| = QAP, \quad ||A|| = QA^*P.$$

If $L = ||A||$ is accepted by an automaton with generating matrix A, consider an automaton with matrix

$$B = \begin{pmatrix} \epsilon I & \epsilon I \\ 0I & A + PQ \end{pmatrix},$$

where I is the unit matrix of the proper order. Then B^* has the element

$$(A + PQ)^* = A^*(PQA^*)^*$$

in the upper left corner. The sum of the elements of the first row of this matrix is $QA^*(PQA^*)^*P$. This sum in turn is a sum of elements

$$(QA^*P)(QA^*P)^k = (QA^*P)^{k+1} = L^{k+1}.$$

Hence the total sum is $L^\wedge$. Adding to this term the term ϵ coming from the first row of ϵI, we see that our automaton accepts L^*.

We can now prove the theorem using the language of series. A rational series is obtained from the empty word by a finite sequence of the following operations on rational series,
1) summing two series
2) multiplying two series
3) starring a series.

One of the terms of 1) may just be a letter. If we proceed by induction after the number of operations performed, we may asssume that every term, factor and series to be dealt with is accepted by an automaton. By the first

11.4 Every rational language is accepted by a finite automaton

part of the proof, the resulting sum, product or starred series is then also accepted by an automaton. This finishes the proof.

R. Let l be the number of signs $+, *, \times$ which are used to define a language L. Prove by induction that there is an automaton with at most $2l$ states which accepts L.

Literature

The original paper by Kleene (1957) dealt with nerve nets including some input neurons. He also took time into account. The arrangement of Kleene's proof presented here is taken from Kuich and Saloma (1985).

References

Aho A.V., Hopcroft J.E., Ullman J.D.(1974): The Design and Analysis of Computer Algorithms. Addison-Wesley 1974.

Auslander L., Tolimieri R.(1979): Is computing with the finite Fourier transform pure or applied mathematics? Bull. AMS 1.2 (1979),847-897.

Auslander L., Feig E., Winograd S.(1984): The Multiplicative Complexity of the Discrete Fourier Transform. Advances Appl. Math. 5.1(1984), 87-109.

Blum M., Micali S.(1987): How to Generate Cryptographically Strong Systems of Pseudo-random Bits. SIAM J. on Computing 13.4, 1984, 850-864.

Burnside W.: The Theory of Groups of Finite Order. Cambridge 1897.

Boole G: The Laws of Thought. Dover Publications.

Cooley J. W., Tukey J.W.(1965): An algorithm for the machine calculation of complex Fourier series. Math. Comput. 19(1965),297-301.

Coppersmith D., Winograd S.(1987): Matrix Multiplication in Arithmetic Progression. Proc. 19th ACM Symposium. New York City 1987.

Gauss C.F.(1801): Disquisitiones Arithmeticae. English translation. Springer 1986.

Harrison M.A.(1965): Introduction to Switching and Automata. New York. McGraw-Hill 1965.

Herlestam T.(1979): Algebra med tillämpningar på kryptologi mm. (Swedish). Math.Inst. Lund 1979

Kleene S.C.(1956): Representations of Events in Memory Nets and Finite Automata. Princeton Univ. Studies 1956,3-40.

Knuth D.(1981): The Art of Computer Programming 2, second edition. Addison-Wesley 1981.

Kuich W., Saloma A.(1986): Semirings, Automata, Languages. Springer 1985.

Lachud G.(1986): Les codes géométriques de Goppa. Astérisque 133/34 (1986)189-206.

Laksov D.(1986): Algebraisk komplexitetsteori. (Norwegian). NORMAT 34.1 (1986).

Macwilliams F.J., Sloane N.J.A.(1977): The Theory of Error-correcting Codes. North-Holland 1977.

Pan V.(1984): How to Multiply Matrices Faster. Springer Lecture Notes in Computer Science. vol. 179 (1984).

Polya G.(937): Kombinatorische Untersuchungen für Gruppen, Graphen und chemische Verbindungen. Acta Math. 68 (1937),145-264.

Riesel H(1985): Prime Numbers and Computer Methods for Factorization. Birkhäuser 1985.

Rivest R. L., Shamir A., Adleman L.(1978): A Method for Obtaining Digital Signatures and Public-Key Cryptosystems. Comm. ACM 21, (1978),120-126.

Solovay R., Strassen V.(1977): A fast Monte-Carlo Test for Primality. Siam J. Comput. 6(1977),84-85.

Schönhage A., Strassen V.(1971): Schnelle Multiplikation grosser Zahlen. Computing 7(1971)281-282

Turing A.M.(1937): On computable numbers with an application to the Entscheidungsproblem. Proc. London Math. Soc. ser. 2, 42(1936),230-265. Corrections ibid. 43 (1937), 544-546.

Winograd S.(1980): Arithmetic Complexity of Computations. CBMS-NSF Reg. Conf.Series in Appl. Math. Philadelphia 1980. Conf. Series in Appl. Math. Philadelphia 1980.

Young A.(1929): On Quantitative Substitutional Analysis (fifth paper). Proc. London Math. Soc. 31(2),(1929),273-288.

Index

abstract
  addition 34
  division 35
  linear algebra 84
  operations 34
  multiplication 35
  subtraction 35
addition 34
Aho-Hopcroft-Ullman 25,33,69
algebraic
  complexity theory 91
  field 36,107
  number 1,16
algorithm
  additive 102
  bilinear 99
  Euclid's 3
  for FFT 63
  Schönhage-Strassen 23,66,69
  general 95
  quadratic 99
  matrix multiplication 97,106
Auslander-Feig-Winograd 69
Auslander-Tolimieri 69
automaton 186

bases
  for vector spaces 85
  for natural numbers 4
bilinear forms 88
binomial theorem 108
Blum-Micali 32
Boole 168
Boolean
  algebra 168-175
  function 173
  polynomial 170
  ring 171
bound
  Singleton 132
  sphere-packing 137
  Gilbert-Varshamov 138

  Plotkin 139
Bose-Chaudhauri-
  Hoquenghem 135
Burnside 167
Burnside's lemma 162

center of a group 142
characteristic 73
Chinese remainder theorem 6,25
class formula 151
code 130
  BCH 135
  block 130
  cyclic 132
  Goppa 140
  linear 131
  quadratic residue 134
  Reed-Solomon 135
complexity of multiplication 24,68
congruence class 5
congruences 5
conjugacy class 147
Cooley-Tukey 69
coordinates 86
Coppersmith-Winograd 106
coprime 3
coset 147
cost of
  addition 21
  subtraction 21
  multiplication 22,24,68
  division 23
  reciprocal 23
  FFT 61
cycle 154
  notation 178
cyclotomic polynomial 121

degree 93,109
derivative 119
dimension

of vector space 85
direct product of
  groups 42, 143
  rings 72
direct sum of
  modules 42
  rings
Dirichlet 2
divide 1
divisibility 1,107,109
  theorem 110
division 35
  ring 36
divisor 1,109
dual space 86

Euclid 1
Euclidean domain 112
Euclid's algorithm 4,6,112
Euler-Fermat theorem 7
Euler's function 7,9

FFT 60
factoring 25
  of large numbers 28
Fermat 7,8,16
Fermat's theorem 8
field 36,75
  automorphisms of 124
  finite 123
  Galois 123
  splitting 117
finitely generated 51,92
Fourier transform
  finite 54
  fast 53,60
formal series 184
formal power series 91
Frobenius 167

Galois 125

Gauss 12,14,18,20,113
Gaussian integers 14
Goppa 140
groups 36, 141-167
  acting on sets 145
  alternating 154
  and combinatorics 162-167, 175-181
  center of 142
  cyclic 143
  dihedral 152
  finite 150-162
  generator of 143
  morphism 148
  of bijections 144
  of low order 158-162
  solvable 157
  symmetric 154

Hamming 130,133
Harrison 181
Herlestam 140
homogeneous 94
Horner 25

ideal 79
image 43,81
indeterminate 92
integer 1
  algebraic 17
integral domain 77
inverse 35
invertible 74

Jacobi symbol 13,26,27

kernel 43,81
Kleene 182,190
Klein 42
Knuth 28,33
Kuich-Saloma 191

Lachud 140
Laksov 106
language 184
  rational 185
Legendre 12
Lehmer 26
linear
  algebra 84-90
  feedback 127
  form 86
  independence 84
  map 89

Macwilliams-Sloane 140
matrix
  rings 73
  multiplication 97,106
  transition 187
  with elements in a non-
    commutative algebra 182-183
Moebius 9,10,150
Moebius's inversion formula 10
modular FFT 64
module 2, 36, 34-52
  Gaussian 14
  cyclic 39
  direct sums of 42
  exponents, types 49
  finitely generated 51
  free 51
  morphisms of 43,81
  over a ring 78
  quotients of 41
monic 109
monoid 36,92,183
  free 92
morphism
  group 148
  module 43
  ring 80
multiplication 35
  complexity 95-101

numbers
  natural 1
  complex 14
  algebraic 16

octogon 153
orbit 146
opposite 35
order 10
  maximal 10,18,19,47
  of a set 150

Pan 106
Pless 140
Polya 167
partial fraction 114
Pepin 26
period 126
  least 126
Plotkin 139
polynomial
  rings 71,91,107
  degree 93
  cyclotomic 121
$Pr(m)$ 10
primary 3
prime 1
  Gaussian 14, 16
  polynomial 109,111
  in characteristic p 120
primitive elements 18
product 35,96
  interior 132
pseudo-random numbers 29
Public Key 28

quadratic reciprocity 12,14
  and finite Fourier transform 58
quotients of
  groups 148
  modules 41

rings 79
vector spaces 89

Reed-Solomon 137
Riesel 28
ring 36
  commutative 70
  division 75
  finite 73
  of matrices 73
  of quotients 79
  morphism 80
  polynomial 71
  subring 70
Rivest-Shamir-Adleman 28

Schönhage-Strassen 23,66,69
shift register 127
sign of a permutation 157
Singleton 132
Solovay-Strassen 26,27
splitting field 117
stabilizer 146
structure
  finite modules 48
subgroup 142
  normal 145
submodule 39
subring 70
subtraction 35
switching function 165, 175
symmetry group 152

trapdoor 28
Turing 182

unit 35

van der Monde 136
Varshamov-Gilbert 138,140
Wilson's theorem 8

Winograd 106

Young 181

zero divisors 74
zeros
  of polynomial mod prime 7
  multiple 119